T0330996

Pacific Northwest Coastal Environments

Pacific Northwest Coastal Environments: Earthquakes and Sea Level Rise investigates the potential impacts of changes in global sea level by examining historical sea and land levels, projected future levels, and by determining how those changes may affect future tides and storm surges to inform their potential to cause harmful impacts. This region has a unique interaction of land, sea, and tectonics. Climate and tectonic change can initiate issues ranging from an inundation of coastal areas due to a combination of sea level rise, vertical land movement, and potential tsunami. This combination of factors leads to the retreat of coastal shorelines due to erosion caused by both tidal action and wave runup. Specific topics explored in this book include the following:

- Coastal erosion rates along the Pacific Northwest coastline from Eureka, California to Vancouver Island, British Columbia.
- Sea cliff retreat mitigation techniques looking at both the advantages and disadvantages of the different techniques.
- Interaction between subduction zone earthquakes and vertical land movement.
- Wave characterization in both deep and shallow water. In addition, the book looks at both wave refraction and reflection along the coastline.
- Tides along the Pacific Northwest coastline and their role in calculating the relative sea level and its effect on coastal erosion.

Pacific Northwest Coastal Environments
Earthquakes and Sea Level Rise

Ronald C. Chaney

CRC Press
Taylor & Francis Group
Boca Raton London New York

CRC Press is an imprint of the
Taylor & Francis Group, an **informa** business

Designed cover image: Shutterstock

First edition published 2025
by CRC Press
2385 NW Executive Center Drive, Suite 320, Boca Raton FL 33431

and by CRC Press
4 Park Square, Milton Park, Abingdon, Oxon, OX14 4RN

CRC Press is an imprint of Taylor & Francis Group, LLC

© 2025 Ronald C. Chaney

ISBN: 978-1-032-59329-6 (hbk)
ISBN: 978-1-032-59330-2 (pbk)
ISBN: 978-1-003-45421-2 (ebk)

DOI: 10.1201/9781003454212

Typeset in Times
by KnowledgeWorks Global Ltd.

This book is dedicated to the following individuals who, in many ways, influenced and helped with the writing of this book.

To my wife

Patricia Jane Chaney

and to

Dr. Kenneth L. Lee

Contents

PART I   *Pacific Northwest Coastal Geology Issues*

PART II   *Northern Pacific Ocean Issues*

PART III   *Coastal Soil and Rock Materials*

PART IV   *Changing Sea Level and Cliff Retreat*

PART V   Effects on Man-Made Structures

About the Author

Ronald C. Chaney is an Emeritus Professor of Environmental Resources Engineering, former Director of the Telonicher Marine Laboratory, and Head of Vessel Operations at Cal Poly Humboldt (formally Humboldt State University) in Arcata, California. Prior to that, he was an associate professor of civil engineering at Lehigh University in Bethlehem, Pennsylvania. He received his BSCE and MSCE from Cal State University-Long Beach and a PhD in engineering from the University of California-Los Angeles. Dr. Chaney was previously the editor of the *International Journal of Marine Geotechnology*, co-editor of the *Journal of Marine Georesources and Geotechnology*, and co-editor of the American Society for Testing and Materials' *Geotechnical Testing Journal*. He is the author and co-author of the books *Seafloor Processes and Geotechnology*; *Environmental Geotechnology*, 2nd edition; and *Marine Geology and Geotechnology of the South China Sea and Taiwan Strait*. He is a fellow of both ASCE and ASTM. Dr. Chaney is also both a licensed civil and geotechnical engineer in California and a retired licensed professional engineer in Oregon.

Preface

The primary purpose of this book is to provide a broad synthesis of concepts intended to serve students, teachers, professional engineers, and geologists on subduction zone earthquakes, tsunamis, ocean wave environment, and the resulting coastal inundation and erosion. The book is designed to serve as a bridge between the standard soil mechanics curriculum of civil engineering and journal articles/books of defined scope in coastal engineering. The author believes that a book encompassing the Pacific Northwest coastal environment needs to broadly cover the areas of interest such as geology, ocean currents and tides, wave characterization, sediment material properties, subduction zone earthquakes, vertical land movement, coastal erosion, inundation, and methods to possibly mitigate their effects. Curriculum development on this subject should be consistent with its multidisciplinary character, a fact that requires serious consideration in the development of any educational program on the subject.

Ronald C. Chaney
Emeritus Professor
Cal Poly Humboldt

Acknowledgments

Many individuals have helped, in a number of ways, over the years. The following are a few of these individuals whom the author wishes to thank: Dr. Kenneth R. Demars, Dr. Donald G. Anderson, Dr. H.Y. Fang, Dr. Adrian Richards, Dr. Armand J. Silva, Dr. Vincent P. Drnevich, Dr. Homa J. Lee, Dr. Willam R. Bryant, Dr. George H. Keller, Dr. Richard H. Bennett, Dr. Kathryn Moran, Dr. Umesh Dayal, Richard S. Ladd, Willard L. DeGroff, Ronald J. Ebelhar, Dr. Lori Dengler, Dr. Harvey Kelsey, Dr. Nitin Pandit, Dr. C.B. Crouse, and Sandra Slonim.

1 Introduction

1.1 THE PURPOSE

The purpose of this work is to look at the effects on the Pacific Northwest coast of the interaction between the Cascadia Subduction Zone (CSZ) and sea-level rise. Sea-level rise is linked to changes in the Earth's climate. A warming climate causes global sea level to rise principally by a number of actions (1) warming the oceans, which causes sea water to expand, increasing ocean volume, and (2) melting land ice, which transfers water to the ocean. However, at regional levels, sea-level rise is affected by a number of additional factors. On the U.S. Pacific Northwest coast, factors include regional climate patterns such as El Niño, which warms the cool Pacific Ocean; the rising and sinking of land along the coast as a result of geologic processes such as plate tectonics, and proximity to Alaska glaciers, which exert a gravitational pull on sea water. Figure 1.1 presents the observed and projected sea-level rise. A review of Figure 1.1 shows that after following an extended time of relative stability, global sea level has begun rising since the late 19th or early 20th century, when global temperatures begun to increase. The most comprehensive assessments of global sea-level rise come from the Intergovernmental Panel on Climate Change (IPCC). Based

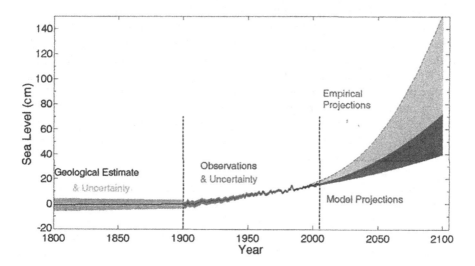

FIGURE 1.1 Observed and projected sea-level rise.

(From National Academy of Science, 2012; Used with permission Board on Earth Sciences and Resources National Research Council (U.S.) Ocean Studies Board from Sea Level Rise for the Coasts of California, Oregon, and Washington: Past, Present, and Future, 2012)

DOI: 10.1201/9781003454212-1

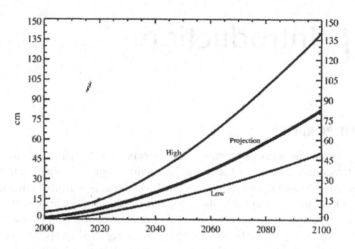

FIGURE 1.2   Sea-level projections.

(From National Academy of Science, 2012; Used with permission Board on Earth Sciences and Resources National Research Council (U.S.) Ocean Studies Board from Sea Level Rise for the Coasts of California, Oregon, and Washington: Past, Present, and Future, 2012)

on tide gage measurements from around the world, the IPCC estimated that global sea level rose an average of approximately 1.7 mm/yr over the 20th century. Since approximately 2000, precise satellite altimetry measurements and tide gage records show that the rate of sea-level rise has increased to approximately 3.1 mm/yr. This rate of sea-level rise is projected to increase at an even faster rate in the future. The high, low, and average projection of sea-level rise as a function of time is presented in Figure 1.2. The interaction of land, sea, and tectonics during a time of climate change involves a wide series of issues that need to be addressed.

This book looks at the potential impacts of changes in global sea level along the U.S. Pacific Northwest coast. This topic is approached by looking at historical sea and land levels, projected future levels, and how those changes may impact future tides and storm surges leading to loss of coastal land. This information will be useful in presenting the potential effects on the coastal communities of the Pacific Northwest. A schematic sketch of major geologic and physiographic features of the continental margin along the Pacific Northwest is presented in Figure 1.3. A review of Figure 1.3 illustrates the complex geologic environment of this region. This topic is approached by looking at the tectonics and resulting vertical land movement, potential inundation of coastal land; currents along the Pacific Northwest coast, tides, wave run-up, role of the geologic materials comprising the coast, coastal retreat, and sea cliff retreat mitigation. In addition, how those changes may impact future tides and storm surges and their potential impact on coastal inundation and land use.

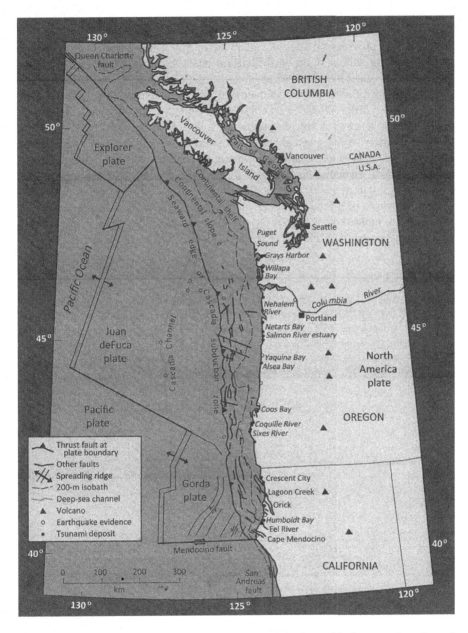

FIGURE 1.3 Schematic sketch of major geologic and physiographic features of continental margin of the Pacific Northwest.

(Atwater and Hemphill-Haley, 1997)

1.2 SUMMARY

Plate tectonics provides a means to explain the overall features and major processes that occur on the surface of the earth. This includes the character of the world's continental margins and its coastal settings.

Plate tectonics allows us to understand why coastal regions on either side of a continent can be so different. The west coast of the United States is bordered by mountains and experiences volcanic eruptions and frequent earthquakes. In contrast, the tectonically inactive east coast abuts a flat coastal plain and is fronted by an almost uninterrupted chain of barriers. The drainage of a continent and how much sediment is delivered to a particular coast is also shown to be dependent on the regions tectonic history combined with climatic factors. Even the location of individual major rivers is shown to be a function of plate tectonic processes. The result is shown that coastlines are molded by numerous physical, chemical, and biological processes, but all these operate on the backbone of the plate tectonic setting of the region.

REFERENCES

Atwater, B., and E. Hemphill-Haley. (1997). Recurrence intervals for great earthquakes of the past 3,500 YEARS at Northeastern Willapa Bay, Washington. US Geological Survey Professional Paper 1576. US Government Printing Office, Washington, DC, National Oceanic and Atmospheric Agency (NOAA), pp. 108.

National Academy of Science. (2012). Sea-level rise for the coasts of California, Oregon, and Washington: Past, Present, and Future. http://dels.nas.edu/besr

Part I

Pacific Northwest Coastal Geology Issues

2 Marine and Coastal Geology

2.1 INTRODUCTION

Initially it is believed that all the continents were once joined together in a super continent called Pangea. This super continent was surrounded by a single world ocean called Panthalassa (i.e., Paleo-Pacific). During the Palezoic-Mesozoic Transition (250 Ma) it occupied almost 70% of the Earth's surface. In the early Jurassic (about 190 million years ago) the Pacific Plate originated from a triple junction formed between the Panthalassic Farallon, Phoenix, and Izanagi Plates (Figure 2.1). The Farallon Plate moved eastward and begun subducting under the west coast of the North American Plate which was located in what is now modern Utah. The Farallon Plate begun sinking at approximately a <30 degrees angle until it was beneath the North America Plate where it then scraped along the bottom of the continent for 1500 km before sinking again into the mantel.

Subduction of a plate at the point of a convergent boundary typically occurs at an angle ranging from 25 to 75 degrees to Earth's surface. Shallower angles of

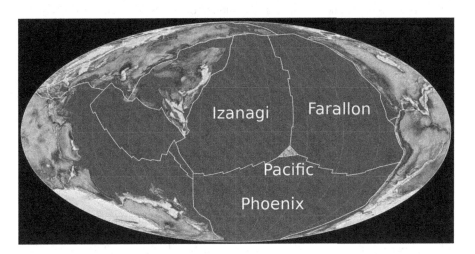

FIGURE 2.1 Schematic depiction of the opening of the Farallon, Phoenix, and Izanagi Plates triple junction forming the Pacific Plate.

(From Wikipedia, Creative Commons CC0 License)

DOI: 10.1201/9781003454212-3

subduction (i.e., flat slab subduction) are also known to exist as well as some that are extremely steep (i.e., steep angle subduction).

- Flat slab subduction (subducting angle <30 degrees) occurs when the slab subducts nearly horizontally. The relatively flat slab can extend for hundreds of kilometers.
- Steep angle subduction (subducting angle >70 degrees) occurs in subduction zones where Earth's oceanic crust and lithosphere are old and thick and have lost buoyancy.

Steep angle subduction is associated with back arc extension of crust, creating volcanic arcs and pulling fragments of continental crust away from continents to leave behind a marginal sea (Stevenson and Turner, 1977; Wikipedia, 2022).

The Farallon Plate acted as a conveyor belt, conveying terranes as it moved eastward. These terranes consisted of a combination of old island arcs and fragments of continental crustal material from other plates. North Americas west coast is primarily made up of these accreted terranes. The Farallon Plate has subducted beneath the North American Plate. This occurs because it was made of basalt material which is denser than the granite which comprised the continental plate.

A new subduction zone is formed when the gravitation tug on the older basalt oceanic crust and uppermost mantle eventually exceeds the bearing capacity of the lower mantle. At this point it begins to break apart and sink. The physical expression of the process is a depression or trench. It is believed that the Cascadia Subduction Zone (CSZ) has not formed a trench yet because it is too young.

The process of subduction reacts in a combination of stress, friction, and chemical changes in the mantel material. This results in the majority of the world's volcanoes and earthquakes. Subduction zones also evolve with time as the geometry, composition, and age of the descending slab change. The changes occur when either a bit of continental crust or a ridge hits the subduction zone.

Continental crust is too light when compared to basalt to be taken into the subduction zone. If the chunk of crust is small such as an island arcs, it gets attached onto a continental margin in a process called accretion. Much of the current geography of the U.S. West Coast is due to this process.

The results of a spreading ridge meeting a subduction zone are also relevant. The first issue is there is no strength across the two sides of a spreading center. The second issue is that once the ridge meets the subduction zone, it dies and ceases to exist.

As a result of the above the giant subduction zone system that once spanned the entire west coast of both North and South America has been cut into segments as different parts of the ridge made contact with the trench (Figure 2.2).

2.2 TECTONICS

The convergent boundary between the Juan de Fuca plate system and the North American plate marks the CSZ. The CSZ is the largest remaining section of the former west coast subduction system along the North American coast. This boundary

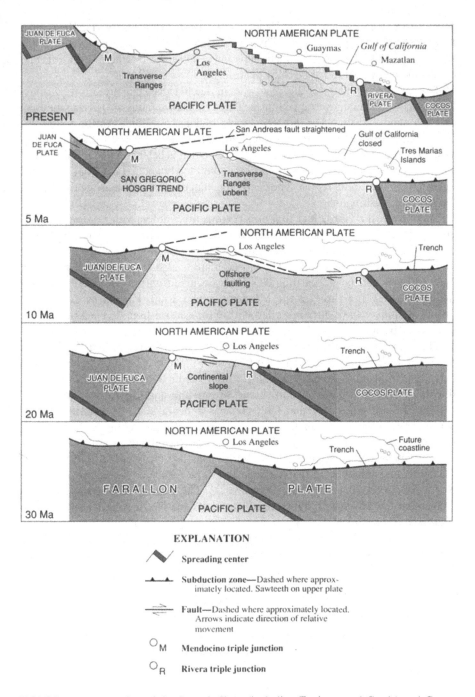

FIGURE 2.2 Formation of the Juan de Fuca (including Explorer and Gorda) and Cocos plates (including Riveria) and of the San Andreas Fault.

(From Wallace, 1990; Courtesy of the U.S. Geological Survey; https://pubs.usgs.gov/pp/1990/1515/pp1515.pdf)

extends from Cape Mendocino to Vancouver Island, Canada, and west to the Gorda, Juan de Fuca, and Explorer ridges. These ridges in the future will also meet the trench and cease to exist. The plate geometry existing at present in Northern California and the Pacific Northwest illustrates how plates cease to exist.

The western boundary of this plate system due to ridges that are offset by transform faults gives an uneven appearance. These transform faults also divide the Juan de Fuca plate into three distinct geographic regions: (1) Gorda (in the south); (2) Juan de Fuca (the central and largest region); and (3) the Explorer (the smallest segment offshore of Canada). This is the only region that is behaving like a classic plate with the majority of earthquakes concentrated on the boundaries and not the plates interior as in the Gorda region. The central or Juan de Fuca region is the offshore of Oregon and Washington.

In contrast, the Gorda and Explorer regions are riddled with intraplate earthquakes, refer to Figure 2.3. As an example, within the Explorer Plate since 2000 over 200 earthquakes of magnitude 4 (M4) and larger have occurred. This includes nine in the M6 range. The Gorda plate is also earthquake prone. More than 130 M4 and larger earthquakes have occurred within the plate's interior including a 7.2 and three in the M6 range since 2000. In contrast, only nine intraplate quakes have been recorded within the central Juan de Fuca in over the same time period.

The Gorda region differs from the Juan de Fuca in Oregon and Washington in several other important ways. The CSZ fault plane is a gently dipping surface extending from the edge of the point of initial subduction at the edge of the continental shelf for 100 to 130 km inland toward the east. In Humboldt County, perhaps a large portion of that fault plane is beneath land. By central Oregon, it is entirely offshore. Second, the rate of plate convergence along the Southern Cascadia margin is less than further north. The Gorda Plate experiences significant intraplate deformation inside its boundaries as discussed earlier. A map of the Gorda and neighboring Juan de Fuca and Explorer Plates subducting under the North American Plate is shown in Figure 2.3. The easterly side of the plate is a convergent boundary with the Gorda Plate under the North American Plate. The southerly side of this plate is a transform boundary with the Pacific Plate along the Mendocino Fault. The westerly side of the Gorda Plate is a divergent boundary with the Pacific Plate forming the Gorda Ridge. This ridge provides morphological evidence of differing spreading rates, with the northern portion of the ridge being narrow, and the southern portion being wide. The northerly side of the Gorda Plate is a transform boundary with the Juan de Fuca Plate, the Blanco Fracture Zone. Numerous faults have been mapped in both the sediments and basement rock of the Gorda Basin. The basin is in the interior of the Gorda Plate south of 41.6 N. Stresses from the neighboring North American Plate and the Pacific Plate cause frequent earthquakes in the interior of the plate. Earthquake epicenters are shown in Figure 2.3.

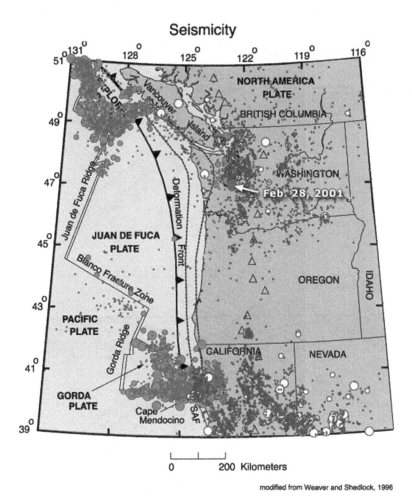

FIGURE 2.3   Epicenter locations in Farallon Plate.

(Courtesy of the Wikipedia Creative Commons CC0 License. https://en.wikipedia.org/wiki/Juan_de_Fuca_Plate#/media/File:Juan_de_fuca_plate.png)

2.3   VERTICAL LAND MOVEMENT

2.3.1   INTRODUCTION

Vertical land movement (VLM) is the movement of land due to the removal of or addition of loads to the upper surface of the earth. VLM is evaluated using a combination of tide gages, benchmark levels, and global positioning systems (GPS) observations.

The Gorda plate subducts beneath the North America plate at approximately 36 mm/yr to form the CSZ fault (McCaffrey et al., 2007; Figure 2.3). The plate deforms elastically, causing vertical land-level change when the fault is seismogenically

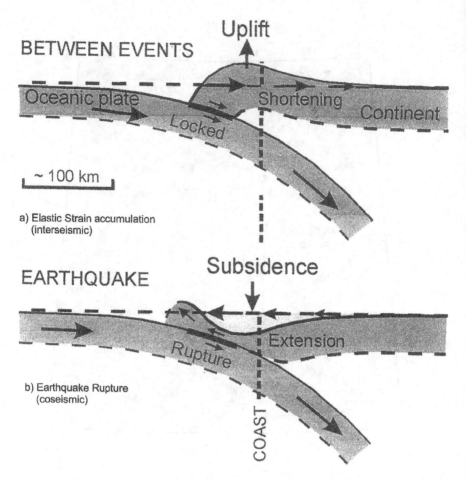

FIGURE 2.4 Vertical motion of ground surface (a) Interseismic, (b) Coseismic.

(Adapted from Hyndman and Wang, 1995; Plafker, 1972; Shugar et al., 2014)

locked (Flück et al., 1997; Mitchell et al., 1994; Wang et al., 2003a). Areas landward of the locked region of the fault generally uplift during the interseismic period, as observed in Japan (Hyndman and Wang, 1995; Loveless and Meade, 2010) and elsewhere (Feng et al., 2012; Wang et al., 2001; Wang and Trehu, 2016; Figure 2.4a). Areas directly above the formally locked region of the fault after a seismic event generally subside during the interseismic period (Figure 2.4b).

2.3.2 Measurement by Region

Local sea-level change is a sum of the vertical change based on sea-level rise and vertical land-level changes (Figure 2.5; Burgette et al., 2009; Nelson and Shennan, 1996b).

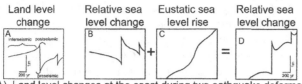

| Land level change | Relative sea level change | Eustatic sea level rise | Relative sea level change |

A) Land level changes at the coast during two eathquake deformation cycles with different amplitude
B) Relative sea level (RSL) changes produced by the cycles
during a period of no change in regional sea level
C) A gradual rise in regional sea level during the cycles that does
not include short term or small scale changes in local and regional
sea level
D) RSL changes at the coast resulting from the sum of figures B
and C

FIGURE 2.5   Tectonic land level (as measured by relative sea level) combined with relative
sea-level change and eustatic sea-level rise to result in the water-level observations recorded
by tide gages.

(Modified from Nelson et al., 1996. Courtesy of the USGS)

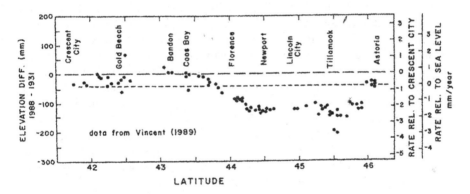

FIGURE 2.6   The rate of the vertical land movement (VLM) compared to the rising global
(eustatic) sea level, based on tide gauge measurements at Crescent City and Astoria.

Notes: (1) Land is rising faster that the eustatic sea level; (2) Land is being submerged by
rising sea level.

**(Adapted from Vincent, 1989, Master's Thesis, modified by Komar and Shih, 1993.
Courtesy of the University of Oregon)**

A graph of the difference in the elevation of coastal land versus latitude for the
coastline from Crescent City, California to Astoria, Oregon was constructed by
Vincent (1989). Vincent compared surveys made in 1931 and 1988 to evaluate elevation changes in the coastal zone, refer to Figure 2.6. The elevation changes shown in
Figure 2.6 are relative rather than absolute. The elevation changes were normalized
using Crescent City, California as the datum. Therefore, the elevation change scale

at the left of the Figure 2.6 gives 0 for Crescent City. Positive values on the elevation change scale indicated at other locations represent an increase in elevation of the land relative to Crescent City. Negative values indicate reduced elevations. If there is a tectonic change in elevation at Crescent City then that would change the elevation accordingly. The smallest uplift of the land occurred along the central coast between Newport and Tillamook is shown in Figure 2.6. Higher uplifts occurred south of Coos Bay and along the very northern portion of the coast toward Astoria and the Columbia river. The general uplift of the Northwest coast can be evaluated by reviewing tide gauge records (Hicks et al., 1985). Based on this review the relative sea-level change measured by tide gauges at Astoria, Oregon is −0.1 to −0.2 mm/yr. In contrast, the rate at Neah Bay on the north coast of Washington State is −1.3 mm/yr, and a Crescent City in Northern California the rate is −0.7 mm/yr. The negative signs indicate that the water level is dropping relative to the land.

The interseismic plate tectonic land-level change associated with the CSZ is characterized as a major plate boundary fault system in the Pacific Northwest of the United States (Chaytor et al., 2004); refer to Figure 2.3.

The northward migrating dextral shear associated with the Pacific North America San Andreas plate boundary fault system is also a source of tectonic deformation (Williams et al., 2002). Understanding this tectonic deformation will enable us to quantify and predict future sea-level trends in Northern California. Sea-level rise at the Humboldt Bay North Spit (NS) tide gage is much greater than any other gage in the Pacific Northwest (Figure 2.7). These NS gage records led some previous researchers to discard these data as apparently anomalous, possibly due to localized site settlement (Mitchell et al., 1994; Verdonck, 2006). National Oceanic and Atmospheric Administration Center for Operational Oceanographic Products and Services (NOAA, 2013) reports an observed sea-level rate of 4.7 mm/yr at the NS tide gage in Humboldt Bay (Figure 2.7). However, other researchers include these data in their models of tectonic deformation (Figure 2.7; modified from Wang et al., 2003b). Tidal models for Humboldt Bay were found to underpredict tidal elevations by ~1 m when NS tidal data were used as an input to their tidal circulation model (Northern Hydrology and Engineering, 2015).

Sea-level rise in the Pacific Northwest has been estimated to be 2.28 mm/yr (Burgette et al., 2009) and 2.38 mm/yr (Zervas et al., 2013). Based on satellite altimetry, global estimates of sea-level rise range up to 3.4 mm/yr (Cazenave and Llovel, 2010; Nerem et al., 2010; Wenzel and Schröter, 2014). The discrepancy between regional sea-level rise estimates and the NS tide gage observations suggests that there is subsidence of the land and the associated tide gage. At the next nearest NOAA continuous operating tide gage in Crescent City CCL California, sea level is observed to be lowering at 0.65 mm/yr (Zervas et al., 2013), this is the result of upward VLM in Crescent City. When the NS tide gage was installed, 11 tidal benchmarks and associated temporary gaging stations were deployed from 1977 to 1980. Utilizing a subset of these initial observation points, analysis of contemporary sea-level observations in Humboldt Bay was used to investigate local sea-level rise relative to regional sea-level rise. This was accomplished using first-order leveling data collected by the National Geodetic Survey (NGS) to determine VLM rates for the second half of the 20th century (Burgette et al., 2013). GPS observations were then incorporated into a continuous analysis of VLM for the past decade. Relative sea-level rise rate for the coastline from Crescent City, CA to Willapa Bay, Washington State is presented in Figure 2.8 (Griggs et al., 2017). A review of Figure 2.8 indicates that it agrees with Figure 2.6 (Komar et al., 2011).

2.4 ESTIMATION OF CASCADIA SUBDUCTION ZONE MAGNITUDE, ATTENUATION, AND RECURRENCE

2.4.1 INTRODUCTION

Ground motion relations for earthquakes that occur in subduction zones are an important input to seismic hazard analyses in the Cascadia region (Atwater and Hemphill-Haley, 1997). This region is comprised of the coastal zones of Northern California, Oregon, Washington State and British Columbia. Along this coastal zone there is a significant hazard from both megathrust and large crustal earthquakes (Nelson et al. 2006, 2014; Rau, 1973). These earthquakes occur along the subduction zone interface and from large events within the subducting slab. These seismic hazards are in addition to the hazard from shallow earthquakes in the overlying crust. In the following the attenuation, magnitude, and recurrence interval will be discussed.

2.4.2 ATTENUATION

An extensive ground motion data base has been compiled by various authors for earthquakes occurring in subduction zones considered representative of the CSZ in the Pacific Northwest (Atkinson and Boore, 2003; Crouse, 1991; Crouse et al., 1988; Youngs et al., 1997).

2.4.3 MAGNITUDE

The magnitude capability of a fault is a function of the length of rupture during a seismic event and its type. In the case of the CSZ it is believed by some authors that the seismic event should be considered in three separate scenarios. A summary of the various scenarios is presented in Table 2.1. These scenarios are the southern half rupturing, or separately the northern half rupturing. The third scenario is for the full length of the fault to rupture. The southern half is considered running from Cape Mendocino, California to approximately Newport, Oregon. The northern half runs from approximately Newport, Oregon to Vancouver Island in British Columbia. There seems to be some agreement in the literature that both the southern and northern halves are capable of generating a moment magnitude of 8 to 8.6. In contrast, if the entire fault would rupture then a magnitude 9+ is possible.

2.4.4 RECURRENCE INTERVAL

The estimation of the recurrence interval between seismic events has a greater range of opinions. As shown in Table 2.1, the northern half of the CSZ has a recurrence interval ranging from 350 to 430 years. In contrast, the southern half of the CSZ

TABLE 2.1
Cascadia Subduction Zone Parameters Assuming Northern and Southern Half

Estimation of Cascadia Seismic Zone Magnitude Length of Rupture	Magnitude	Occurrence Probability in 50 Years	Recurrence Interval (Years)
• **Southern half** (1)	8–8.6	25–40	240
• Newport, Or. to Coos Bary (4)			300–380
• Coos Bay, Or. to Eureka (4)			220–240
• **Northern half** (1)		10	350
• Central and Northern Oregon (4)		14–17	430
• Washington/British Columbia (4)		8–14	350
• Newport to Astoria (4)			430
• Astoria to Vancouver Island (4)			
• **Full** (1)	9	14	480
• (3)	7.1	37	50
• (2)	≥8.8		500
	≥9.0		1000
	≥9.3		10,000

Notes: (1) Thompson (2011); (2) Rong et al. (2014) based on magnitude and recurrence rate; (3) Oregon Department of Emergency Management; (4) Today.oregonstate.edu.

has a recurrence interval ranging from 220 to 380 years. This can be broken down further for section from Coos Bay to Eureka having a recurrence interval from 220 to 240 years. For a seismic event involving a full rupture of the fault ranges from 50 to 10,000 years. The 50-year estimate by the Oregon Department of Emergency Management can probably be discarded as an outlier. The estimates ranging from 500 to 10,000 years by Rong et al. (2014) are based on magnitude. The larger magnitudes (i.e., 8.8 to 9.3) having a larger recurrence time. A review of Table 2.1 indicates that the southern half is expected to more frequent earthquakes but at a lessor magnitude (probability of occurrence in 50 years ranging from 25% to 40%) than the northern half (probability of occurrence in 50 years ranging from 10% to 17%).

2.5 COASTAL AND MARINE GEOLOGY

2.5.1 INTRODUCTION

The Pacific Northwest coastline extends from Northern California (i.e., Cape Mendocino to the Oregon border), Oregon and Washington State. The PNW coastline encompasses a wide range of coastal landforms. These coastal landforms include sandy and cobble beaches backed by coastal dunes, coastal bluffs, or steep cliffs. This coastline also includes uplifted terraces, barrier spits, and estuaries and lagoons. A summary of regional geomorphic coastal features is presented in Table 2.1.

In this section, the morphology of each of the cells comprising the coastline will be discussed. The littoral cells 1 through 3 are of interest for the California coast in the present study; Oregon in Figures 2.12 and 2.13; and Washington State in Figures 2.13 and 2.14.

2.5.2 COASTAL NORTHERN CALIFORNIA

The coastline of Northern California is often inaccessible, and includes rocky cliffs and hills with the occasional long sandy beaches. The coastline extending from Centerville Beach near Ferndale to the mouth of the Klamath River is mostly beach accessible. The littoral cells along the California coast are present in Figure 2.8. For the purpose of this study only the first 3 cells are of concern. The coastline itself is composed of uplifted, terraced, and wave-cut thick Mesozoic and Cenozoic sedimentary strata.

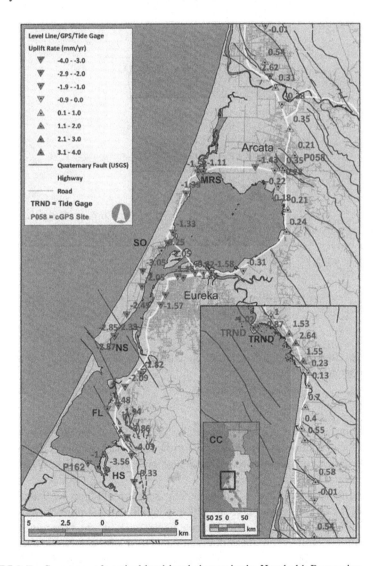

FIGURE 2.7 Summary of vertical land-level change in the Humboldt Bay region.

(From Patton et al., 2017)

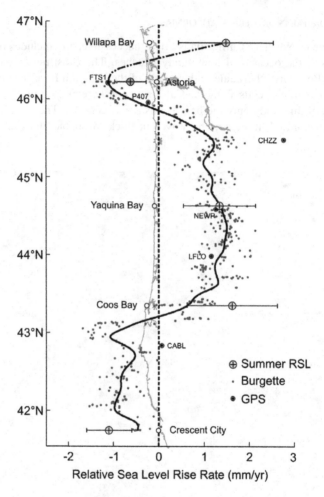

FIGURE 2.8   Relative sea-level rise rate for the coastline from Crescent City, CA, to Wallapa Bay, Washington State.

(From Griggs et al., 2017)

In addition, there are a number of smaller bays and coastal lagoons created by sand spitz located along the Northern California coast. An example of this process is Humboldt Gay. It consists of three distinct segments: (1) Arcata Bay, (2) South Bay, and (3) Entrance Bay.

The sediment distribution of Humboldt Bay is shown in Figures 2.9 and 2.10.

Humboldt Bay is a coastal lagoon, 22 km long and a maximum of 6.5 km wide. The bays are divided into three distinct segments, each of which occupies the seaward end of one or more stream valleys cut into Plio-Pleistocene paralic sediments. The segments are linked to narrow tidal channels which are bounded by a barrier spit on the west and by high valley interfluves on the east. Tides are of the mixed type with a mean range of 2 m.

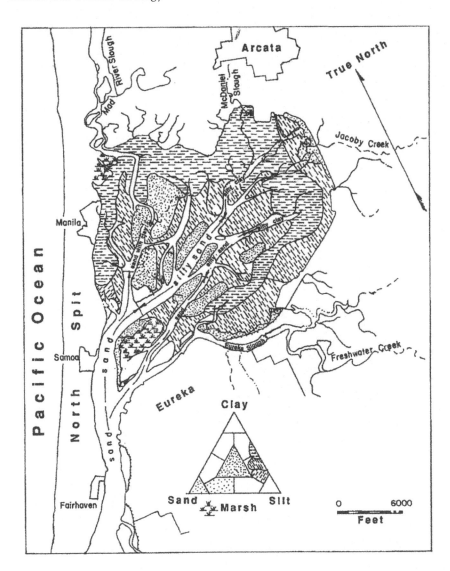

FIGURE 2.9   Sediment distribution in Arcata Bay.

(From Thompson, 1971. Courtesy of Humboldt State University)

Textural variations of the surface sediments in the South and Arcata bays corre-late generally with bay-floor morphology. Bottom sediments of the inward branching tidal channels are covered by gravelly, shelly sand which becomes finer and more muddy with increasing distance from the tidal influence at the bay entrance.

Big Lagoon is a brackish embayment at the mouth of Maple Creek, refer to Figure 2.11. It is the second largest of four coastal lagoons located in Humboldt County, California (Porter, 1984). It is located approximately 40 km north of Humboldt Bay. The lagoon is separated from the ocean by a narrow sand spit

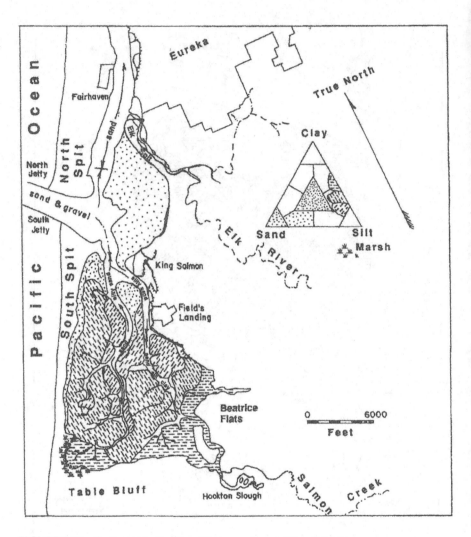

FIGURE 2.10   Sediment in South and Entrance Bays of Humboldt Bay.

(From Thompson, 1971. Courtesy of Humboldt State University)

which breaches to the ocean in most years. Breaching of the lagoon occurs at
the north end of the lagoon typically during December-March each year due to
increased river runoff from Maple, Tom, and Pitcher Creeks (Brady, 1977). Joseph
(1958) describes the lagoon as 5.6 km long and 2 km wide with the greatest depth
varying between 7.6 m and 10.4 m, depending on the season.

Brady (1977) recorded four physical cycles occurring in Big Lagoon on an annual
basis related to the breaching of the sand spit. The pre-breached period was characterized as being isohaline (intermediate salinity) with increased precipitation until the
sand spit breached. During the breached period the lagoon is under tidal influence.
The water column experiences mixing resulting in high salinities. The post-breached

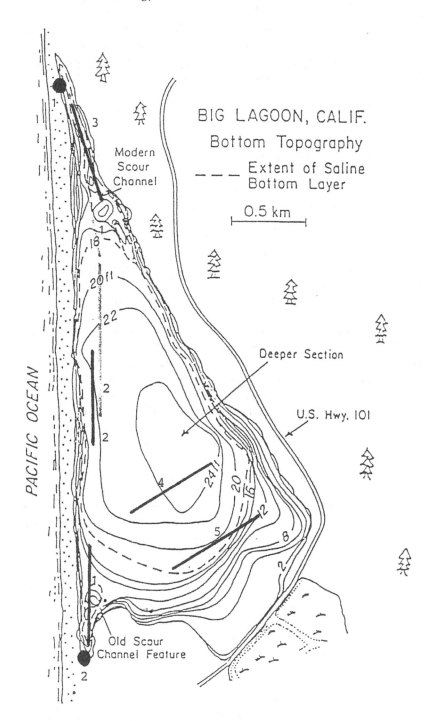

FIGURE 2.11   Big Lagoon, California Bottom Topography.

(From Magenheim and Pequegnat, 1986. Courtesy of Humboldt State University)

period is initiated with the closing of the sand spit. A two-layered system develops at this time with dense saline water on the bottom separated from the less saline surface layer by a halocline. No mixing occurs as spring progresses between the two layers and the bottom compartment eventually becomes oxygen depleted.

2.5.3 OREGON COASTLINE

2.5.3.1 Introduction

In general, the southern part of the coast and the northern most part near the Columbia River are rising faster than the present rate of sea-level rise, while the north central stretch has minimal uplift and therefore is experiencing sea-level transgression due to the global rise in sea level. This coast wide pattern of relative sea-level change is reflected in the degree of cliff erosion.

The Klamath Mountains in Southern Oregon extend to the coast, and characterized by high rocky cliffs and small pocket beaches. The north central stretch has minimal uplift and therefore is experiencing a sea-level transgression due to the global rise in sea level.

Extensive stretches of beach are found in the lower lying parts of the coast extending from Coos Bay to Tillamock.

Erosion has been common along the Oregon coast due to the high energy of the wave climate and the dynamic behavior of its beaches. Among these areas is the Oregon Dunes which is north of Bandon, the largest complex of dunes in the Continental United States. In addition, the longest continuous beach extends from Coos Bay northward to Heceta Head near Florence, a total shoreline length of approximately 100 km. Along the northern part of the coast there is an interplay between sandy beaches and rocky shores. Massive headlands are present and between them are stretches of sandy shoreline whose lengths are governed by the spacings between the headlands.

On a coast-wide scale, much of the difference in the degree of bluff erosion reflects the pattern of coastal tectonic uplift versus the global rise in sea level (Habel and Armstrong, 1978; Hapke and Reid 2007; Komar and Shih, 1993; Komar et al., 2011). On the southern half of the coast, the tectonic uplift presently exceeds the global rise in sea level, and cliff erosion there has been minimal within historic times, although there is evidence for substantial bluff retreat in the past. Along the north-central coast between Newport and Tillamook, global sea-level rise exceeds the tectonic uplift (Komar and Shih, 1993); sea cliff erosion in that area has been more significant.

Erosion occurs primarily in a series of littoral cells formed by large headlands which effectively isolate the stretches of beach within each cell. Sources and losses of sand to the series of littoral cells are highly variable and this has controlled the amount of sand on the beach and the elevation of the beach/cliff junction. In the following each of the littoral cells will be discussed.

2.5.3.2 Southern Oregon Littoral Cell

Southern Oregon littoral subcell exists between the California border and Cape Blanco. This section of the coastline contains six littoral subcells as shown in Figure 2.12.

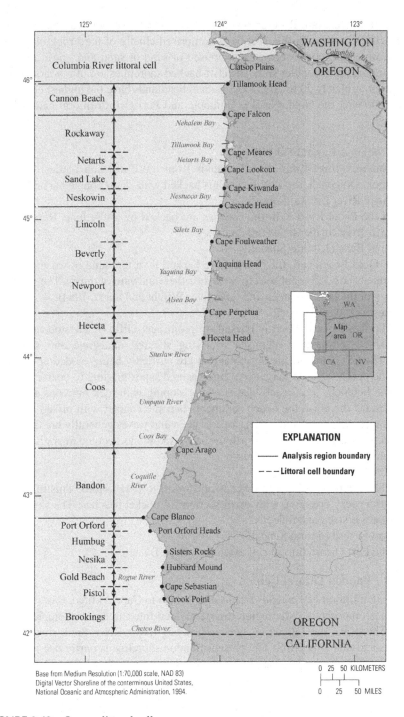

FIGURE 2.12 Oregon littoral cell.

(From Ruggiero et al., 2013)

The littoral cells are Brookings, Pistol, Gold Beach, Nesika, Humbug, and Port Orford. The coastline in this section is comprised chiefly of hard rock bluffs with periodically small pocket beaches composed of sand or sand and gravel mix with clasts ranging up to boulder size. The cliff-backed shorelines are of moderately high relief (approximately 20 to 30 m). Occasional headlands have the highest elevations. In the following the Gold, Nesika, Humbug, and Port Orford littoral cells will be discussed.

1. Brookings Littoral Subcell

   The Brookings Littoral Cell extends from the California border to Saddle Rock, Oregon. The coastline is bluff backed with small cobble barrier beach.

2. Pistol River Littoral Subcell

   In the Pistol River cell, the beaches are backed by a foredune. Remnant sea cliffs exist further landward on dunes.

3. Gold Beach Littoral Subcell

   In Gold Beach cell, the beaches are backed by a foredune of varying spatial dimensions. Remnant sea cliffs exist further landward of dunes. The morphology of the beach is steep and reflective (Wright and Short, 1984).

4. Nesika Littoral Subcell

   The Nesika cell beach is backed by prominent cliffs of Pleistocene marine terrace sedimentary deposits overlying sheared mudstone and sandstone rocks (i.e., Jurassic age deposits) that are currently being eroded at the base. The beach deposits range from fine-grained sands grading to coarser cobbles and boulders. The beach face is gently sloping with rock outcrops and reefs. Toward the north the beach sediments become coarser with mixed sand and gravel. As a result the beach slope is steep and waves generally break directly on the beach face. Further north the beach sediments become finer grained and the slopes of the beach decrease.

5. Humbug Littoral Subcell

   The Humbug littoral cell extends northward to Port Orford. This littoral cell has a rocky coastline with small sections of pocket beaches. The slope of the beach is steep and reflective.

6. Port Orford Littoral Subcell

   The Port Orford littoral cell extends to Cape Blanco.

2.5.3.3 Bandon Littoral Cell

The Bandon littoral cell extends from Cape Blanco to Cape Akrago. Based on geology and geomorphology the region can be divided into three morphological beach types. These are the following: barrier beaches (New River Spit), dune-backed beaches, and plunging cliffs. The present Bandon shoreline is partly due to coastal tectonics. In this case the coast is emergent relative to sea-level rise. The offshore sea stacks are probably a relict of the CSZ earthquake in 1700. In this event the shore subsided and extensive erosion occurred. Beach sediment grain size decreases toward the north. Aeolian processes have moved fine-grained sand inland and formed dunes. Sand foredunes have also formed at the back of the beach along the length of the New River Spit.

2.5.3.4   Coos Littoral Cell

1. The Coos Littoral subcell
   The Coos littoral subcell extends from Heceta Head to Cape Arago. The Coos littoral cell is primarily made up of wide, straight beaches punctuated by three large estuaries and associated river mouths – Coos Bay, Winchester Bay, and the Siuslaw River Estuary. The rivers associated with these estuaries are the following: Siuslaw, Umpqua, Coos, and Millacoma rivers mouth(s). The Coos Bay dune sheet is the largest dune accumulation in the United States, extending from Coos Bay to *Heceta Head*. The cell is a dune and bluff-backed shoreline.
2. Heceta Littoral Subcell
   The Heceta littoral cell extends from Heceta Head to Cape Perpetua. It is relatively small, little more than 10 km, and not populated. Rocky coastline with headlines and coves.

2.5.3.5   Lincoln County Region Littoral Cells

The Lincoln County region extends between Cape Perpetua and Cascade Head. The region contains three littoral cells. These cells are the following: Newport, Beverly, and Lincoln.

1. Newport Subcell
   The Newport cell contains two estuaries: Alsea Bay and Yaquina Bay. In the vicinity of Alsea Bay and Yaquina Bay, the beaches are backed by dunes. The remainder of the cell is backed by coastal bluffs and marine terraces.
2. Beverly Subcell
   The Beverly littoral cell extends from Yaquina Head to Cape Foulweather. It is sand starved because the only source of modern sand is the local erosion of the bluffs backing the beach. The bluffs consist mainly of tertiary mudstones capped by a thin layer of Pleistocene terrace sands.
3. Lincoln Subcell
   North of Cape Foulweather and a stretch of rocky coast to Depoe Bay the Lincoln cell extends to Cascade Head. Siletz Bay is fronted by Siletz Spit barrier spit. The remainder of the beaches in this cell are backed by sea cliffs which are the main source of sand to the beach. This sand is coarse grain sand which mixes with medium-grained sand on the beach. The beaches fronting the Siletz Spit and Bleneden beach have the coarsest sand with sizes decreasing to both the north and south. The coarse-grained beaches are relatively steep and intermediate to reflective. Beaches along Lincoln City are low in slope and dissipative.

2.5.3.6   Tillamook County Region Littoral Cells

The Tillamook County region extends from Cascade Head to Cape Falcon. It is made up of long sandy beaches interspersed with prominent headlands and steep bluffs. This region contains four littoral subcells: Neskowin, Sand Lake, Netarts, and Rockaway. Tillamook County coast is submergent relative to water-level changes.

1. Neskowin Subcell
   The Neskowin subcell extends from Cascade Head to Cape Kiwanda. It is made up of long sandy beaches interspersed with prominent headlands and steep bluffs.
2. Sand Lake Subcell
   The Sand Lake subcell extends from Cape Kiwanda to Cape Lookout. It is made up of long sandy beaches interspersed with prominent headlands and steep bluffs.
3. The Neskowin subcell
   The Neskowin subcell extends from Cape Lookout to Cape Meares. It is made up of long sandy beaches interspersed with prominent headlands and steep bluffs.
4. Rockaway Subcell
   The Rockaway subcell extends from Cape Meares to Cape Falcon. It is made up of long sandy beaches interspersed with prominent headlands and steep bluffs.

2.5.3.7 Cannon Beach Region Littoral Cells

The Cannon Beach region extends from Cape Falcon to Tllamook Head. This cell consists of a combination of wide sandy beaches backed by broad dunes, and cobble and boulder beaches backed by bluffs. Gravels are prevalent along much of the littoral cell shoreline forming a thin veneer at the back of most of the beaches in southern Cannon Beach. Geomorphically many beaches may be characterized as composite because the beaches consist of a wide dissipative sandy beach, backed by a steep upper foreshore of gravels. In addition, several of the bluff-backed sections have well-vegetated faces. This littoral cell is at a transition between areas with positive and negative relative sea-level rates.

2.5.3.8 Columbia River Littoral Cell

Clatsop Plains subcell
The Clatsop Plains subcell is an arcuate-shaped coastline that extends from Tillamook Head in the south to the mouth of the Columbia River. It is part of the Columbia River littoral cell. The Clatsop Plains subcell coastline is characterized by wide, dissipative surf zones and prominent longshore bars in the nearshore, whereas the beaches are backed by an extensive dune sequence.

2.5.4 WASHINGTON STATE COASTLINE

2.5.4.1 Introduction

The coast of Washington State is characterized by river and alpine glacier sediments above basalt and marine sedimentary rocks that were accreted to the continent. The southern coastline is lined with sandy sediments that work their way from mouths of rivers. The coast was formed through a process controlled by a combination of the CSZ and accretion of material. North of the town of IIwaco at the mouth of the Columbia River, Willapu Bay and Gray's Harbor exist. These two lagoon areas are two

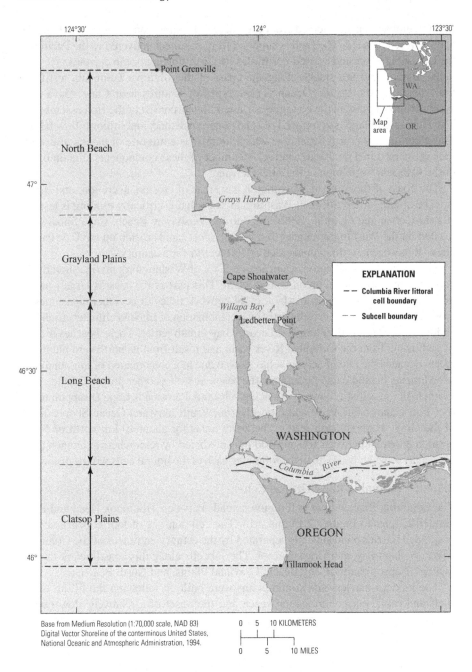

FIGURE 2.13 Columbia River littoral cell.

(From Ruggiero et al., 2013)

shallow embayments protected from the Pacific Ocean by long low ridges that have been covered by sand. The sand is supplied from sediment delivered to the Pacific by the Columbia River and carried northward by north-flowing ocean currents.

Headlands and sea stacks along the coast of the Olympic Peninsula testify to the erosion by the Pacific Ocean. These types of features near Cape Alava (i.e., Westward most point in the contiguous United States) consist of the Hoh assemblage. The Hoh assemblage is composed of sandstone, mudstone, and minor pillow basalt. These components form a mélange which literally is a mixture of blocks of diverse rock immersed in a mudstone matrix. A number of these components contain basalt of the Crescent Formation of the Siletz terrane.

The coast of the State of Washington has areas of erosion, accretion, and stability. The amount of coastline in Washington State that is critically eroding is less 1% (Bernd-Cohen and Gordon, 1999). In contrast, Washaway Beach, Cape Shoalwater located on the Washington State coast has the most rapid erosion on the U.S. Pacific coast. It has been eroding an average of 100 feet/yr for a century.

The southwest Washington ocean coast (~2.4% of Washington's marine shorelines), is backed by dunes and a broad coastal plain. This part of the coast is largely accretional with pockets of erosion which cycle between times of accretion and times of erosion. This behavior is driven partly by a combination of El Niño climate variability, and/or complications resulting from nearby bay mouth jetties. These beaches are fed by sediments from the Columbia River basin and result from its northward plume. In recent decades the rate of accretion has slowed due to a combination of impoundment of sediments behind dams throughout the basin, as well as other potential factors. An example of areas that is now experiencing substantial erosion is Cape Disappointment (i.e., Fort Canby State Park) and Grays Harbor South Jetty and Ocean Shores (north of the Grays Harbor North Jetty). The shore areas for about 10 km north of North Head may retreat between 100 and 300 m is predicted by researchers at Oregon State University to occur by 2020. In the following each of the littoral cells will be discussed.

2.5.4.2 Columbia River Littoral Cell

The Columbia River littoral cell region extends between Tillamook Head and Point Grenville, refer to Figures 2.13 and 2.14. The cell consists of four concave-shaped prograded barrier plain subcells separated by the estuary entrances of the Columbia River, Willapa Bay, and Grays Harbor. The subcells along this coastline are the following: Clapsop Plains, Long Beach, Grayland Plains, and North Beach.

The modern barriers and strand plains were built up following the filling of the shelf and estuary accommodation space and the onset of a relatively slow rate of eustatic sea-level rise about 6,000 years ago. About 4,500 years ago, Long Beach and Clatsop Plains begin to prograde, whereas Grayland Plains begin to prograde about 2800 years ago. Wide, gently sloping beaches characterize the modally dissipative high-energy system. The beaches are backed predominantly by prograded dune fields, and swales and by sea cliffs along the northern half of the North Beach subcell.

    1. Long Beach Subcell and Grayland Plains Subcell
       The Long Beach and Grayland subcells have broad surf zones with multiple sandbars characterizing the modally dissipative high-energy system.

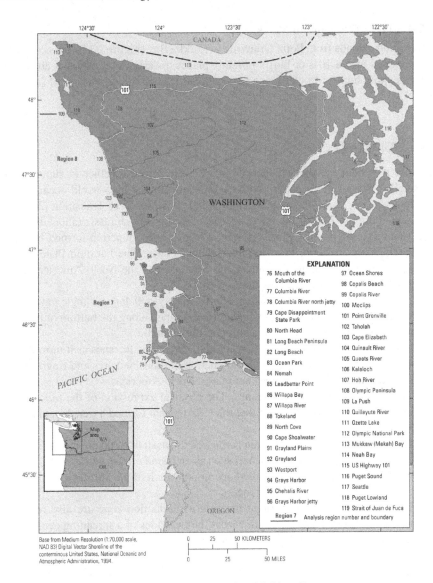

FIGURE 2.14 Coastline between Cape Elizabeth to Mukkaw Bay.

The beaches are backed predominantly by prograded dune fields and swales and by sea cliffs or bluffs along the northern half of the North Beach subcell.

2. North Beach Subcell

The North beach subcell extends from Grays Harbor to Point Greville. Beach lopes in the subcell are steepest at the Southern profiles and decrease to the north. Median grain sizes are largest near the mouth of Grays Harbor and decrease to the north. Dune crest elevations are highest at the southern end of the subcell and decrease in height to the north. North of the Copalis River, the beaches are backed by cliffs or bluffs.

2.5.4.3 Point Grenville to Mukka Bay

This region extends from Point Grenville to Mukka Bay (Figure 2.14). This coastline is composed of seacliffs or bluffs in many places towering over 150 feet above the beach, expose both bedrock of Hoh rocks and large continuous outcrops of sand and gravel.

2.6 SUMMARY

Initially it is believed that all the continents were joined together in super continent called Pangea. This super continent was surrounded by a world ocean called Panthalassa (i.e., Paleo-Pacific). During the Palezoic-Mesozoic Transition (250 Ma) it occupied almost 70% of the Earth's surface. In the early Jurassic (about 190 million years ago) the Pacific Plate originated from a triple junction formed between the Panthalassic Farallon, Phoenix, and Izanagi Plates. The Farallon Plate moved eastward and begun subducting under the west coast of the North American Plate which was located in what is now modern Utah.

The Farallon Plate begun sinking at approximately <30 degrees angle until it was beneath the North America Plate where it then scraped along the bottom of the continent for 1500 km before sinking again into the mantle.

The Farallon Plate acted as a conveyor belt, conveying terranes as it moved eastward. These terranes consisted of a combination of old island arcs and fragments of continental crustal material from other plates. North Americas west coast is primarily made up of these accreted terranes. The physical expression of the process is a depression or trench. It is believed that the CSZ has not formed a trench yet because it is too young.

Continental crust is too light when compared to basalt to be taken into the subduction zone. If the chunk of crust is small such as an island arcs, it gets *attached* onto a continental margin in a process called accretion. Much of the current geography of the U.S. West Coast is due to this process.

The results of a spreading ridge meeting a subduction zone are also relevant. The first issue is there is no strength across the two sides of a spreading center. The second issue is that once the ridge meets the subduction zone, it dies and ceases to exist.

As a result of the above the giant subduction zone system that once spanned the entire west coast of both North and South America has been cut into segments as different parts of the ridge made contact with the trench. The convergent boundary between the Juan de Fuca plate system and the North American plate is the result of this process and marks the CSZ. The CSZ is the largest remaining section of the former west coast subduction system along the North American coast. This boundary extends from Cape Mendocino to Vancouver Island, Canada, and west to the Gorda, Juan de Fuca, and Explorer ridges. The western boundary of this plate system due to ridges that are offset by transform faults give an uneven appearance. These transform faults also divide the Juan de Fuca plate into three distinct geographic regions: (1) Gorda (in the south), (2) Juan de Fuca (the central and largest region), (3) and the Explorer (the smallest segment offshore of Canada). This is the only region

TABLE 2.2

Regional Geomorphic Coastal Features of Northern California and the Pacific Northwest

Coastal Feature	N. Calif.	S. Oregon	Brandon	Region Name Oregon Coos Bay	Lincoln Country	Tillamook Country	Cannon Beach	Org. and Wash CRLC	Washington Olympic Pen
Rocky shorelines and platforms	X	X	X		X		X		X
Wide and narrow sandy beaches backed by dunes	X		X	X	X	X	X	X	
Gravel and cobble beaches backed by cliffs						X	X		X
Plunging cliffs	X	X	X	X	X	X	X	X	X
Barrier spits	X		X	X	X	X	X	X	X
Estuaries	X	X	X	X	X	X	X	X	X

Source: Adapted from Ruggiero et al. (2013).

that is behaving like a classic plate with the majority of earthquakes concentrated on the boundaries and not the plates interior as in the Gorda region. The central or Juan de Fuca region is offshore of Oregon and Washington. The Gorda plate subducts beneath the North America plate to form the CSZ fault. The CSZ potential earthquake magnitude and recurrence interval considering the fault acting in three scenarios based on length of rupture are also presented. These three scenarios are (1) full rupture, (2) rupture of northern half, and (3) rupture of southern half. The various plates deforms elastically, causing vertical land-level change when the fault is seismogenically locked. Areas landward of the locked region of the fault generally uplift during the interseismic period. Areas directly above the formally locked region of the fault after a seismic event generally subside during the interseismic period. This vertical movement of the land whether upward or downward in addition to the eustatic rise in sea-level results in a changing relative sea level along the northwest shoreline. This changing relative sea level along the shoreline results in potential erosion and cliff retreat.

REFERENCES

Atkinson, G. M., and Boore, D. M. (2003). Empirical ground motion relations for subduction zone earthquakes and their application to Cascadia and other regions. *Bulletin of the Seismological Society of America*, 93(4), 1703–1729.

Atwater, B., and Hemphill-Haley, E. (1997). Recurrence intervals for great earthquakes of the past 3,500 years at Northeastern Willapa Bay, Washington. US Geological Survey Professional Paper 1576. US Government Printing Office, Washington D.C., 108 pp. National Oceanic and Atmospheric Agency (NOAA).

Bernd-Cohen, T., and Gordon, M. (1999). State of the beach/state reports/WA/beach erosion. *Coastal Management*, 27, 187–217.

Brady, C. A. (1977). Yearly cycle of phytoplankton productivity in relation to the hydrography of Big Lagoon. M.S. Thesis, Humboldt State University, Arcata, CA.

Burgette, R. J., Schmidt, D. A., and Weldon, R. J. II. (2009). Interseismic uplift rates for western Oregon and along-strike variation in locking on the Cascadia subduction zone. *Journal of Geophysical Research*, 114, B01408. http://doi.org/10.1029/2008JB005679

Burgette, R. J., Watson, C. S., Church, J. A., White, N. J., Paul Tregoning, P., and Coleman, R. (2013). Characterizing and minimizing the effects of noise in tide gauge time series: Relative and geocentric sea level rise around Australia. *Geophysical Journal International*, 194(2), 719736. https://doi.org/10.1093/gji/ggt131

Cazenave, A., and Llovel, W. (2010). Contemporary sea level rise. *Annual Review of Marine Science*, 2, 145–173.

Chaytor, J. D., Goldfinger, C., Dziak, R. P., and Fox, C. G. (2004). Active deformation of the Gorda plate: Constraining deformation models with new geophysical data. *Geology*, 32(4), 353–356. https://doi.org/10.1130/G20178.2

Crouse, C. B. (1991). Ground-motion attenuation equations for earthquakes on the Cascadia subduction zone. *Earthquake Spectra*, 7(2), 201–236. http://doi.org/10.1193/1.1585626

Crouse, C. B., Vyas, Y. K., and Schell, B. A. (1988). Ground motions from subduction zone earthquakes. *Bulletin Seismological Society of America*, 78, 1–25.

Feng, L., Newman, A. V., Protti, M., González, V., Jiang, Y., and Dixon, T. H. (2012). Active deformation near the Nicoya Peninsula, northwestern Costa Rica, between 1996 and 2010: Interseismic megathrust coupling. *Journal of Geophysical Research*, 117, B06407. http://doi.org/10.1029/2012JB009230

Flück, P., Hyndman, R. D., and Wang, K. (1997). 3-D dislocation model for great earthquakes of the Cascadia subduction zone. *Journal of Geophysical Research*, 102, 20,539–20,550.

Goldfinger, C. (2016). Subduction zone earthquakes off Oregon, Washington more frequent than previous estimates, Oregon State University, Newsroom, August 5: https://today. oregonstate.edu/archives/2016/aug/subduction-zone-earthquakes-oregon-washington-more-frequent-previous-estimates

Griggs, G., Arval, Cayan, D., DeConto, R., Fox, J., Fricker, H. A., Kopp, R. E., Tebaldi, C., and Whiteman, E. A. (2017). California Ocean Protection Council Science Advisory Team Working Group, Rising Seas in California: An Update on Sea-Level Rise Science, California Ocean Science Trust.

Habel, J. S., and Armstrong, G. A. (1978). Assessment and Atlas of Shoreline Erosion Along the California Coast. Sacramento, California, State of California, Department of Navigation and Ocean Development, 69 pp.

Hapke, C. J., and Reid, D. (2007). National assessment of shoreline change, part 4: Historical coastal cliff retreat along the California coast: U.S. Geological Survey Open File Report 2007–1133, 51 pp.

Hicks, S. D. (1985). Tidal datums and their uses - A summary. *Shore and Beach*, 53, 27–33.

Hyndman, R. D., and Wang, K. (1995). The rupture zone of Cascadia great earthquakes from current deformation and thermal regime. *Journal of Geophysical Research Atmospheres*, 1002(B11), 22133–22154. http://doi.org/10.1029/95JB01970

Joseph, J. (1958). Studies of Big Lagoon, Humboldt County. California, M.S. Thesis, Humboldt State University, Arcata, CA.

Komar, P. D., Ruggiero, P., and Allen, J. C. (2011). Sea level variations along the U.S. Pacific Northwest coast: Tectonics and climate controls. *Journal of Coastal Research*, 27(5), 747–765.

Komar, P. D., and Shih, S. -M. (1993). Cliff erosion along the Oregon coast: A tectonic-sea level imprint plus local controls by beach processes. *Journal of Coastal Research*, 9(3), 747–765.

Loveless, J., and Meade, B. (2010). Geodetic imaging of plate motions, slip rates, and partitioning of deformation in Japan. *Journal of Geophysical Research*, 1–35.

Magenheim, J., and Pequegnat, J. E. (1986). Evidence for a shift from nitrification to denitrification in a stratified lagoon, Telonicher Marine Laboratory Report Series 8, Humboldt State University.

McCaffrey, R., Qamar, A. I., King, R. W., Wells, R., Khazaradze, G., Williams, C. A., Stevens, C. W., Vollick, J. J., and Zwick, P. C. (2007). Fault locking, block rotation and crustal deformation in the Pacific Northwest. *Geophysical Journal International*, 169, 1315–1340. https://doi.org/10.1111/j.1365-246X.2007.03371.x

Mitchell, C. E., Vincent, P., Weldon, R. J., and Richards, M. (1994). Present-day vertical deformation of the Cascadia margin, Pacific Northwest, United States. *Journal of Geophysical Research*, 99, 12257–12277.

National Oceanic and Atmospheric Administration (NOAA). (2013). Extreme vertical land motion from long-term tide gauge records. National Ocean Service Center for Operational Oceanographic Products and Services, Silver Spring, MA. NOAA Technical Report NOS CO-OPS 065.

Nelson, A. R., DuRoss, C. B., Witter, R. C., Kelsey, H. C., Engelhart, S. E., Mahan, S. A., Gray, H. J., Hawkes, A. D., Horton, B. P., and Padgett, J. S. (2021). A maximum rupture model for the central and southern Cascadia subduction zone—reassessing ages for coastal evidence of megathrust earthquakes and tsunamis. *Quaternary Science Reviews*, 261(4), 106922.

Nelson, A. R., Kelsey, H. M., and Witter, R. C. (2006). Great earthquakes of varying magnitude at the Cascadia subduction zone. *Quaternary Research*, 65(3), 354–365.

Nelson, A. R., and Personius, S. F. (1996a). The potential for great earthquakes in Oregon and Washington: An overview of recent coastal geologic studies and their bearing on segmentation of Holocene ruptures, Central Cascadia subduction zone. In Rogers, A.M., Walsh, T.J., Kockelman, W. J., Priest, G. R. (Eds.), *Earthquake Hazards in the Pacific Northwest of the United States*. U.S. Geological Survey Professional Paper 1560, 91–114.

Nelson, A. R., Shennan, I., and Long, A. J. (1996b). Identifying coseismic subsidence in tidal-wetland stratigraphic sequences at the Cascadia subduction zone of Western North America. *Journal of Geophysical Research-Solid Earth*, 101, 6115–6135.

Nerem, R. S., Chambers, D. P., Choe, C., and Mitchum, G. T. (2010). Estimating mean sea level change from the TOPEX and Jason altimeter missions. *Marine Geodesy*, 33, 435–446.

Northern Hydrology and Engineering (NHE). (2015). Humboldt Bay: Sea Level Rise, Hydrodynamic Modeling, and Inundation Vulnerability Mapping. Prepared for the State Coastal Conservancy and Coastal Ecosystems Institute of Northern California, McKinleyville, CA. https://humboldtbay.org/sites/humboldtbay2.org/files/Final_HBSLR_Modeling_InundationMapping_Report_150406.pdf

Patton, J. R., Williams, T. B., Anderson, J. K., and Leroy, T. H. (2017). Tectonic and land level changes and their contribution to sea-level rise, Humboldt Bay region, Northern California: 2017 Final Report. Prepared for U.S. Fish and Wildlife Service Coastal Program. Cascadia GeoSciences, McKinleyville, CA.

Plafker, G. (1972). Alaskan earthquake of 1964 and Chilean earthquake of 1960: Implications for arc tectonics. *Journal of Geophysical Research*, 77, 901–925.

Porter, R. S. (1984). Food and feeding of staghorn sculpin (Leptocottus armatus Girard) and the stary flounder (Platichtys stellaatus Pallas) in euryhaline environments. M.S. Thesis, Humboldt State University, Arcata, CA.

Rau, W. W. (1973). Geology of the Washington Coast between Point Grenville and Hoh River, Washington Dept. of Natural Resources, Geology and Earth Resources Division, Bulletin No. 66, 58 pp.

Rong, Y., Jackson, D. D., Magistrale, H., and Goldfinger, C. (2014). Magnitude limits of subduction zone earthquakes. *Bulletin of the Seismological Society of America*, 104(5), 2359–2377.

Ruggiero, P., Kratzmann, M. G., Himmelstoss, E. A., Reid, D., Allan, J., and Kaminsky, G. (2013). National assessment of shoreline change: Historical shoreline change along the Pacific Northwest coast. U.S. Geological Survey Open-File Report 2012–1007, 62 pp. http://doi.org/10.3133/ofr20121007

Shugar, D. H., Walker, I. J., Lian, O. B., Eamer, J., Neudorf, C., McLaren, D., and Fedje, D. (2014). Post-glacial sea-level change along the Pacific coast of North America. *Quaternary Science Reviews*, 97, 170–192.

Stevenson, D. J., and Turner, J. S. (1977). Angle of subduction. *Nature*, 270, 334–336.

Thompson, J. (2011). Cascadia's Fault, Counterpoint, Berkeley.

Thompson, R. W. (1971). Recent sediments of Humboldt Bay, Eureka, California. Humboldt State University, Special Collections Digitized Publication.

Verdonck, D. (2006). Contemporary vertical crustal deformation in Cascadia, *Tectonophysics*, 417(3–4), 221–230, http://doi.org/10.1016/j.tecto.2006.01.006

Vincent, P. (1989). Geodetic Deformation of the Oregon Cascadia Margin: Master's Thesis, University of Oregon.

Wallace, R. E., ed. (1990). The San Andreas fault system, California: U.S. Geological Survey Professional Paper 1515, 283 pp. http://pubs.usgs.gov/pp/1988/1434/.

Wang, K., and Trehu, A. M. (2016). Some outstanding issues in the study of great megathrust earthquakes – The Cascadia example. *Journal of Geodynamics*, Elsevier.

Wang, E., Wan, J., and Liu, J. (2003a). Late Cenozoic geological evolution of the foreland basin bordering the West Kunlun range in Pulu area: Constraints on timing of uplift of Northern margin of the Tibetan Plateau, *Journal of Geophysical Research*, 108(B8), 2401. http://doi.org/10.1029/2002JB001877

Wang, K., Wells, R., Mazzotti, S., Hyndman, R. D., and Sagiya, T. (2003b). A revised dislocation model of interseismic deformation of the Cascadia subduction zone. *Journal of Geophysical Research,* 108, 2026. http://doi.org/10.1029/2001JB001227

Wang, S., Xu, X., and Song, G. (2001). Tectonic features of the foreland thrust zone in the Hotan sag of the southwest Tarim depression. *Petroleum Geology & Experiment*, 23(4), 378–383.

Wenzel, M., and Schröter, J. (2014). Global and regional sea level change during the 20th century. *Journal of Geophysical Research - Oceans*, 119(11), 7493–7508. https://doi.org/10.1002/2014JC009900

Wikipedia. (2022). Subduction. https://en.wikipedia.org/wiki/Subduction

Williams, H., Hutchinson, I., and Nelson, A. (2002). Multiple sources for late Holocene tsunamis at Discovery Bay, Washington State. American Geophysical Union, 2002 fall meeting, abstract S22B-1032.

Wright, L. D., and Short, A. D. (1984). Morphodynamic variability of surf zones and beaches: A synthesis. *Marine Geology*, 56(1–4), 93–118.

Youngs, R. R., Chiou, S. -J., Silva, W. J., and Humphrey, J. R. (1997). Strong ground motion attenuation relationships for subduction zone earthquakes. *Seismological Research Letters*, 68(1), 58–73.

Zervas, C., Gill, S., and Sweet, W. V. (2013). Estimating vertical land motion from long-term tide gauge records. NOAA Tech. Rep. NOS CO-OPS 65, 22 pp.

Part II

Northern Pacific Ocean Issues

3 North Pacific Ocean

3.1 INTRODUCTION

Currents are important agents of coastal erosion. These currents may be the agent of both removal and transport of sediment in the coastal environment. These currents carry sediment that is removed from its resting place by waves. The waves cause the sediment to become temporarily suspended in the water and the currents move it. In the following the cause of currents will be discussed along with Ekman transport and the California Current System.

3.2 CURRENTS

3.2.1 INTRODUCTION

The causes of ocean currents is wind and density differences between water masses. The various types of currents are presented in Figure 3.1. Winds sweeping across the ocean exert a frictional force on the sea surface. This force horizontally displaces

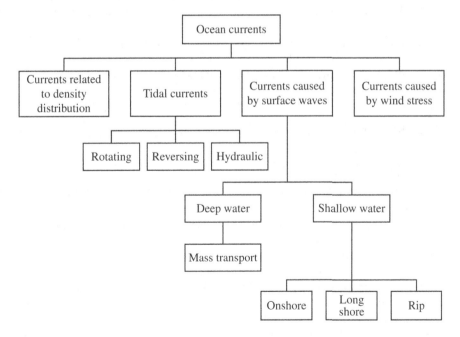

FIGURE 3.1 Types of marine currents.

(From Richards, 1981; republished with permission of Taylor & Francis)

DOI: 10.1201/9781003454212-5

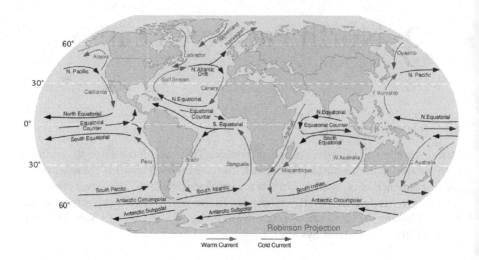

FIGURE 3.2   Ocean currents.

(Courtesy of the Wikipedia Creative Commons CC0 License)

the affected water mass resulting in circulation pattern as shown in Figure 3.2. The speed and direction of currents are modified by a number of factors. These include the Coriolis effect, the shape of ocean basins, thermohaline circulation, and the presence of land masses.

Thermohaline circulation is the result of density differences between water masses. This results in the sinking of surface waters. Thermohaline circulation is thus responsible for a slow overturning of the oceans waters. This mixing recharges the surface waters with nutrients.

3.2.2   Surface Wind-Driven Currents

3.2.2.1   Wind-Driven Circulation

Oceanic phenomena depend upon the physical interaction between the atmosphere and the ocean. Winds in the lower atmosphere are responsible for waves, surface currents, mixing of the surface layer of the ocean, and the exchange of energy in the form of heat and water vapor. Particulate matter ranging from mineral grains, seeds, and spores to carbon and silica spherules are transported out to sea through atmospheric circulation.

Currents due to winds are set in motion by a combination of moving air masses and shaped by the earth's rotation and the configuration of the continents. Moving air affects the sea surface and the waters move in response. The relationship between wind speed and water is not straightforward. Water masses in the upper layers flow at an angle to the wind direction due to the Coriolis effect. Assume initially a stationary sea surface in the northern hemisphere. A gentle breeze begins to blow, gradually intensifying over time. As the wind blows across the sea surface, a column of water, whose depth is a function of wind speed, duration, and fetch moves in response. The column of water does not move uniformly. As the wind exerts a frictional stress on the water surface layer, the water begins to move in the direction of the wind but is immediately acted upon by the Coriolis effect. This deflects the water motion to the right of the wind direction in the northern hemisphere and to the left in the southern hemisphere. Simultaneously, fictional stress between the surface layer and the water layer below it retards the movement of the surface layer. The resultant flow of the surface layer represents a steady state condition in which the surface water moves in a direction 45 degrees to the right of the wind direction in the northern hemisphere or 45 degrees to the left of the wind direction in the southern hemisphere.

The net movement of the water column affected by the wind can be visualized as a series of individual layers of water. By applying the same principles as those presented for the movement of the surface layer, it can be shown that each progressively deeper layer of water moves to the right of the layer above it and at a slightly lower velocity. Depicting the movement of the water layers as a series of vectors whose length corresponds to relative velocity as shown in Figure 3.3. A review of this figure shows that it resembles a spiral staircase with successively shorter steps arranged at progressively greater angles from the wind direction. This circulation pattern is called the Ekman spiral after Swedish physicist V.W. Ekman, who first described it mathematically. The effect of both the deflecting for the earth's rotation (Coriolis effect) and of the eddy viscosity (A) was taken into account by Ekman (1902). The eddy viscosity if defined as follows:

$$\tau_s = A\frac{dv}{dn} \tag{3.1}$$

where

$\frac{dv}{dn}$ is the shear of the observed velocities.

A is the expression for the transfer of momentum of mean motion.

τ_s is the *shearing stress*.

Eddy viscosity, depends upon state of motion of the fluid and is not a characteristic physical property.

Assuming the eddy viscosity is independent of depth (D) then the following expression can be written:

$$\sigma_x = C_1 e^{\frac{\pi}{D}z} \cos\left(\frac{\pi}{D}z + c_1\right) + C_2 e^{\frac{-\pi}{D}z} \cos\left(\frac{\pi}{D}z + c_2\right) \tag{3.2}$$

$$\sigma_y = C_1 e^{\frac{\pi}{D}z} \sin\left(\frac{\pi}{D}z + c_1\right) - C_2 e^{\frac{-\pi}{D}z} \sin\left(\frac{\pi}{D}z + c_2\right) \tag{3.3}$$

$$D = \pi\sqrt{\frac{A}{\rho\omega\sin\varphi}} \tag{3.4}$$

where C_1, C_2, c_1, c_2 are constants that depend on the boundary conditions.

Assuming D is a large number then the motion near the bottom of the water column is zero, then $C_1 = 0$. In addition, assume that the stress (τ_a) is along the y axis, then

$$\tau_a = -A\left(\frac{dV_y}{dz}\right)_0 \tag{3.5}$$

$$0 = A\left(\frac{dV_x}{dz}\right)_0 \tag{3.6}$$

allow C_2 and c_2 to be determined. If V_0 is the velocity at the surface, then

$$V_x = V_0 e^{\frac{-\pi}{D}z}\cos\left(45° - \frac{\pi}{D}z\right) \tag{3.7}$$

$$V_y = V_0 e^{\frac{-\pi}{D}z}\sin\left(45° - \frac{\pi}{D}z\right) \tag{3.8}$$

$$V_0 = \frac{\tau_a}{\sqrt{\rho A2\omega\sin\varphi}} = \frac{\pi\tau_a}{D\rho\omega\sin\varphi\sqrt{2}} \tag{3.9}$$

Therefore, the direction of water movement in the northern hemisphere is 45 degrees to the right of the wind as shown schematically in Figure 3.3. In the southern hemisphere the water motion is to the left of the wind direction.

Under ideal conditions, the Ekman spiral results in a net transport (Ekman transport) of wind-driven water in the affected water column (Ekman Layer) which is

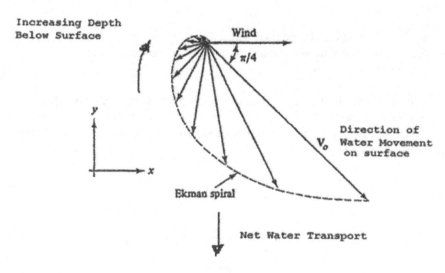

FIGURE 3.3   Net movement of water column affected by wind.

(Adapted from Ekman, 1902)

90 degrees to the right from the wind direction in the northern hemisphere and to the left in the southern hemisphere. In shallow water it will be somewhat less than 90 degrees because of the restricting influence of the ocean bottom.

3.2.2.2 The Oceanic Gyres and Geostrophic Currents

A characteristic feature of the major surface currents is that they describe large circular orbits, or gyres, within individual ocean basins. The circulation of these gyres is anticyclonic (i.e., clockwise in the northern hemisphere and counterclockwise in the southern hemisphere). Typical of the ocean basin gyres is a major clockwise current flow in the North Pacific called the North Pacific gyre. The gyre itself resembles a hill with its peak within the gyre. The hill in oceanic gyres only rise about 2 m above the base of their slopes. Nonetheless, the hills have a significant effect on ocean circulation.

The mounding up of water in oceanic gyres is caused by Ekman transport. The net transport of an entire column of wind-driven water is approximately 90 degrees from the wind direction and about 45 degrees from the direction of the uppermost surface currents as discussed previously. Consequently, there is a continual transport of surface waters toward the interior of a gyre. This convergence of waters results in a slight rise of the sea surface. The mounding up of relatively warm, low-density surface waters in the interior of the gyres depresses the colder, higher-density waters below.

Consider a water element on the slope of this mound. In response to gravity the element begins to move down the mound; however once the element is in motion the Coriolis effect deflects them to the right of its downward path. The balance established between the pressure-gradient force and the Coriolis effect causes the surface waters in the gyre to follow a curved path conforming to the contours of the mound as shown in Figure 3.4 and limits the slope and height of the mound. The net effect is that the water element moves around the mound.

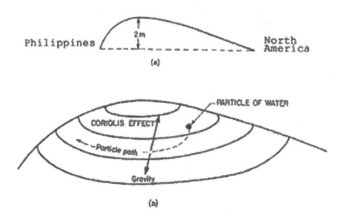

FIGURE 3.4 Coriolis effect on surface water in a gyre.

(Adapted from Parker, 1985)

When the Coriolis effect and the pressure gradient force are in balance, they are said to be in geostrophic balance. The movement of surface waters around an oceanic gyre is therefore called geostrophic currents. The geostrophic currents are responsible for the circular movement of waters in oceanic gyres and conceivably would serve to maintain the gyres for extended period of time even if the global winds suddenly ceased.

3.2.2.3   Westward Intensification of Ocean Currents

The individual currents which comprise a specific oceanic gyre do not flow at the same speeds. Currents along the western margins of ocean basins are considerably swifter and narrower than their counterparts on the eastern side. An example of this effect is the velocity of the Gulf Stream approaches 9 km/hr, while the Canary Current flows south along the European coast at less than 1 km/hr (Parker, 1985). This western intensification of gyre currents results from a balance between several factors which affect the surface currents. These include the increasing magnitude of the Coriolis effect with latitude, latitudinal variations of wind strength and direction, and friction between the water currents and contiguous land masses. These factors cause the topographic highs of the oceanic highs of the gyres to be offset to the west as shown schematically in Figure 3.4. The western boundary currents move faster than those on the eastern side because a constant volume of water is squeezed into a narrower band along the western margins of the ocean basins.

3.2.2.4   Ekman Transport and Vertical Water Movement

Horizontal displacement winds cause substantial vertical displacement of water masses adjacent to continental margins. A phenomenon known as upwelling occurs when wind patterns cause a continuous mixing of water from depths to the surface. A reverse vertical flow called down welling can occur under other wind circumstances. Wind-induced vertical mixing has major effects on biologic productivity and the resulting occurrence of biogenic sediments.

Upwelling occurs next to continental margins and along the equator when prevailing winds cause Ekman transport of surface water away from the affected area. Deep waters then move upward to replace them. In the northern hemisphere this will occur when the coast is to the left of the wind direction and in the southern hemisphere when the shoreline is to the right of the prevailing wind. Deep, cold waters rich in the nutrients which stimulate growth of microscopic floating plants (i.e., phytoplankton) rise up to take the place of the vacating surface waters. Zones where this process occurs are along the coast of Peru, and along the west coast of North American. Equatorial upwelling occurs along the equator, because of the Ekman transport of surface water away from the equator due to westward blowing trade. To take the place of the diverging water masses deep, cold water then rises up. When water is piled up along a coast due to Ekman transport toward shore down welling occurs.

3.2.2.5   Wind-Driven Circulation in the North Pacific

The North Pacific is an example of a large-scale wind-driven circulation pattern. The circulation patterns of the major oceans characteristically involve both a large central gyre and one or more secondary gyres.

The North Pacific general circulation pattern of surface waters is deflected to the right and pushed toward the interior of the mid-Pacific gyre, where a high-pressure zone is maintained. This is due to the trade winds which blow between 30 degrees north latitude and the equator. This movement of surface waters establishes a major, westward flowing geostrophic current, the North Equatorial Current. The current reaches the western margin of the North Pacific Basin, where land masses deflect it toward the north. There it merges with a portion of the Kuroshio Current which has been deflected to the north by the Philippines. The combined current moves along the east coast of the Philippines, Taiwan, and Japan as the Kuroshio Current. The Kuroshio Current then merges with the North Pacific current (NPC) which in turn merges with the California current along the west coast of North America.

When the NPC reaches the western margin of the North Pacific Basin, part of it flows north and then west. Another branch moves into the Bering Sea as the Aleutian Current. The rest flows south along the coast of North America as the California Current. Eventually, the California Current rejoins the North Equatorial Current, completing the closed loop of the North Pacific gyre.

Between the North and South Equatorial Currents flows a narrow, eastward-flowing current called the Equatorial Countercurrent. Other eastward flowing currents move along the equator in the Pacific, Atlantic, and Indian (seasonally) Oceans in addition to the equatorial countercurrents. These are subsurface currents, generically referred to as equatorial undercurrents. An example is the Pacific equatorial undercurrent, also known as the Cromwell Current.

Equatorial undercurrents are approximately 250 m thick and 250 km wide and reach speeds of up to 5 km/hr. The cause of equatorial undercurrents is not completely understood. It has been suggested that the equatorial undercurrent is an eastward return flow of water resulting from the North and South Equatorial Currents. This flow would then be a result of the pressure gradient caused by the slope of the sea surface.

3.2.2.6   Rip Currents

When waves approach a shore diagonally, the water attempts to keep on moving in the same direction. Since the water is stopped by the land, the water particles are deflected along the shore. Therefore, if the waves are approaching from the left, the reflected motion is ordinarily to the right, and vice versa. The resulting flow is called a longshore current. In some cases currents actually move in a direction opposite to that from which the waves are approaching. This results from a convergence that pushes waves together, thus building up the surface waves over a ridge. The result of this convergence is a current that flows out from the point of high level, moving along the shore in a direction that in this case is opposite to that from which the waves are approaching.

As a result of both of these types of longshore currents, considerable scouring out of holes is likely to occur just outside the beach. This results in trough-like area of rather deep water occurring near shore. Further out from the beach usually a bar exists with shallower water.

Note: the energy of the wave crest moving up a submarine canyon is decreased due to divergence (spreading), while the energy on the ridges is increased due to convergence (Shepard, 1969).

The water carried in by the waves must return, otherwise the water would keep getting higher and higher along the shore. A path is formed for the return at some low point in the longshore bar. This return flow, which is particularly pronounced along those beaches where there are large waves, develops a dangerous offshore movement known as rip current (i.e., rip tide). The flows become much stronger in the rip currents but are diminished outside the breaker zone (Shepard, 1969).

3.2.3 CALIFORNIA CURRENT SYSTEM

3.2.3.1 Introduction

The NPC (sometimes referred to as the North Pacific Drift) is a slow water current that flows west-to-east between 30 and 50 degrees north latitude east of the date line in the Pacific Ocean. This is associated with a decrease in the speed of current. The NPC forms the northward portion of the North Pacific Subtropical Gyre, a large swirling current that occupies the northern basin of the Pacific.

Originating from a flow occurring east of the island of Honshu, Japan, the NPC extends over 40 degrees of longitude. The NPC may be considered to be an extension of the Kuroshio Current as shown in Figure 3.2. The current covers a large area and transports warmer water from the subtropics to the sub-polar latitudes.

As the NPC approaches the west coast of North America, it divides into two broad currents: the northward-flowing Alaska Current and the southward-flowing California Current.

The California and the Gulf of Alaska Currents receive different volumes and flows of water from the North Pacific Current. Approximately 60% of the flow of the NPC goes to the Gulf of Alaska. The California Current receives the remaining 40% of the flow. These flow values may fluctuate, resulting in changes in the volume and the speed of the NPC water. The northern ocean current movement southward down the west coast results in much cooler ocean temperatures.

The California current begins off southern British Columbia and ends off southern Baja California. It is considered an eastern boundary current because of the influence on its course of the North American coastline. The California Current flows southward year-round in a zone from the shelf break to a distance of approximately 1000 km from the coast as shown in Figure 3.5 (Hickey, 1979, 1998). The current is strongest at the sea surface, and generally extends over the upper 500 m of the water column.

The Davidson Current is a weak countercurrent, moves warmer saltier water northward during the winter months along the surface within a relatively narrow (10 to 40 km) band. In addition, the California undercurrent (subsurface) flows northward over the continental slope during the summer and early fall as a nearly

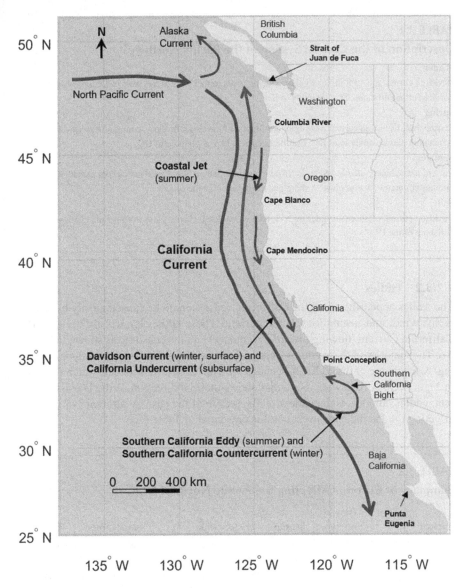

FIGURE 3.5 Major currents in the California current system.

(Wiki California Current System)

continuous feature at depths of about 100 to 400 m beneath the southward-flowing upper layers. The undercurrent has a jet like structure, with the core of the jet located just seaward of and just below the shelf break and with peak speeds of 30 to 50 cm/s. A summary description of the current system of the pacific northwest in presented in Table 3.1.

TABLE 3.1
Description of the Current System of the Pacific Northwest

Winter

Davidson current moves north to south producing counterclockwise eddies west of DC and clockwise eddies further offshore. CC moving N to S.

Spring

Coastal Jest, Davidson current moving N-S with other flow closer to shore, eddies clockwise and counterclockwise offshore of DC, California current N to S offshore of DC.

Summer

Current immediately offshore S_N, California current offshore N-S, Davidson current meandering N-S. Eddies counterclockwise close to shore and clockwise offshore of DC.

Fall

Davidson current offshore meandering N-S, eddies counterclockwise close to shore and clockwise offshore of the DC.

3.2.3.2 Eddies

The eddies appearing along the California coast seem to be caused mostly by topography, wind, and instabilities in the current. These eddies lay mainly between the California Current (flowing toward the equator) and the coastline as shown in Figure 3.6. The majority of these eddies were cyclonic and have the ability to induce upwelling. A schematic of the eddies associated with the California current system is shown in Figure 3.6. A review of this figure shows that the direction of the Davidson current, eddies, and jets is dependent on the season of the year. A summary of currents affecting the Pacific Northwest coast is presented in Table 3.2.

TABLE 3.2
Summary of Currents Affecting the Pacific Northwest Coast

Current	Flow Direction	Season	Location	Velocity (m/s)	Temp (C)
North Pacific	W to E	Winter Summer	30–50 degrees North Latitude	0.0 (5 cm/s)	7.2–16.1 17.8–23.3
California	N to S	Year Round	Shelf break to 1000 m offshore	Strongest at surface to 500 m depth. Seasonal mean speeds ~10 cm/s	cold
Davidson	S to N	Winter Summer	Surface 10–40 km band		
California Under Current	S to N	Year Round	Narrow band, ~10–40 km, seaward and just below shelf break	~30–50 cm/s	
Coastal Jet	-	Year Round	Below the shelf break seaward		

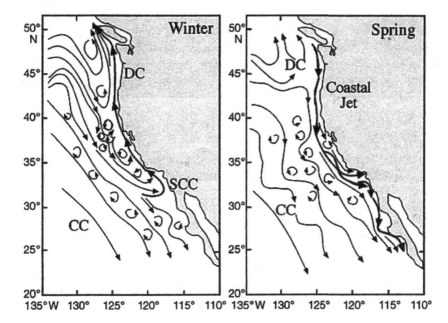

FIGURE 3.6 Schematic of the California Current System eddies.

(Adapted from Hickey et al., 2019)

3.3 WEATHER SYSTEM

3.3.1 EL NIÑO WEATHER SYSTEM

El Niño (i.e., Southern Oscillation, or "ENSO" for short) is a recurring climate pattern across the Pacific Ocean. It is the warming of the ocean surface to above average sea surface temp (SST) in the central and eastern tropical Pacific Ocean. The pattern shifts back and forth irregularly every two to seven years, bringing predictable shifts in ocean surface temperature and disrupting the wind and rainfall patterns across the tropics and the west cost of the United States. Main impacts of temperature and precipitation generally occur in the six months between October and March.

El Niño years are often marked by severe winter storms that bring high levels of precipitation to California's coast. On land and along the coast, this can translate to increased erosion, flooding, and landslides. El Niño winters are associated with higher than normal sea levels that increase the normal erosion cycle.

The pacific northwest coast was mostly characterized by accretion during the past century. Erosion was localized. During the 1997–1998 El Niño winter storms, a series of events (erosion at Cape Shoalwater, Point Brown, and Fort Canby State Park in Washington State) caused erosion along the coast. Subsequently, a return to accretion patterns has resulted in much of that erosion being eliminated.

"The stormy conditions of the 2009-10 El Niño winter eroded the beaches to often unprecedented levels at sites throughout California and vulnerable sites in the Pacific Northwest," said Barnard P. et al. 2011. In the pacific northwest, the regional impacts were moderate, but the southerly shift in storm tracks, typical of El Niño winters, resulted in severe local wave impacts to the north-of-harbor mouths and tidal inlets.

3.4 SUMMARY

Currents are important agents of coastal erosion. These currents may be the agent of both removal and transport of sediment in the coastal environment. These currents carry sediment that is removed from its resting place by waves. The North Pacific Current (NPC; sometimes referred to as the North Pacific Drift) is a slow water current that flows west-to-east between 30 and 50 degrees north latitude east of the date line in the Pacific Ocean. As the NPC approaches the west coast of North America, it divides into two broad currents: the northward-flowing Alaska Current and the southward-flowing California Current. This current generates eddies which cause either permanent or near permanent removal of sediment and rock from the shoreline areas. This process is the result of many actions; some that are slow and small in scale, and others that are very short and intense.

REFERENCES

Barnard, P. L., Allan, J., Hansen, J. E., Kaminsky, G. M., Ruggiero, P., and Doria, A. (2011). The impact of the 2009-10 El Niño Modoki on U.S. West Coast beaches. *Geophysical Research Letters*, 38(13).

Ekman, V. W. (1902). On the influence of the Earth's rotation on ocean-currents. *Arkiv for Matematik, Astronomi Och Fysik*, Band 2(11), 52 pp.

Hickey, B. M. (1979). The California current system: Hypotheses and facts. *Progress in Oceanography*, 8(4), 191–279.

Hickey, B. M. (1998). Coastal oceanography of Western North America from the tip of Baja California to Vancouver Island. In Brink, K. H., and Robinson, A. R. (Eds.), *The Sea, Volume 11*, pp. 345–393. New York, NY: Wiley and Sons, Inc.

Hickey, B. M., Royer, T. C., and Amos, C. M. (2019). California and Alaska currents. In Cochran, J. K., Bokuniewicz, H. J., and Yager, P. L. (Eds.), *Encyclopedia of Ocean Sciences, Volume 3* (3rd ed.), pp. 318–329. Academic Press.

Parker, H. S. (1985). *Exploring the Oceans: An Introduction for the Traveler and Amateur Naturalist*. Engelwood Cliffs, NJ: Prentice-Hall.

Richards, A. F. (1981). Personal communication.

Shepard, F. P. (1969). *Submarine Geology* (3rd ed.). Harper and Row.

4 Tides

4.1 INTRODUCTION

The tide is the periodic rising and falling of the level of the sea caused by the effect of gravitational attraction of the moon and sun on the rotating earth. Tides follow the moon's rotation (i.e., lunar) more closely than they do the movement of the sun. The lunar day (i.e., rotation of the moon around the earth) is about 50 minutes longer than the solar day, therefore tides occur on the average 50 minutes later each day. In most cases there are usually two high and two low waters in a tidal or lunar day. Because of the varying gravitational effects of the sun and moon, a diurnal inequality in the height of tides occurs. This difference is between the high water and its following low water that occurs in a day when compared but against the following days high water and its succeeding low water. Along the Pacific coast tides compare in height with those on the Atlantic coast but have a decided diurnal inequality. Typical tidal ranges along the Pacific seacoast are given in Table 4.1.

Tides experience cyclic occurrence of high and low water along our seashores in response to the gravitational attraction of the sun and moon. The term tide refers only to the relatively short period, astronomically induced vertical change in the height of the sea surface. This term excludes wind-generated waves and swells.

This motion is modified by the rotation of the earth, friction forces, and the ocean boundaries. Additional factors such as coastline configuration, local depth of the water, sea-floor bathymetry, and other hydrographic and meteorological influences may also play a role in changing the tidal range, variation between high and low water, and time of arrival of the tides. A summary of tidal datum is presented in Figure 4.1.

TABLE 4.1
Typical Tidal Ranges

Local		Approximate Ranges (ft.)	
From	To	Mean	Diurnal
Point Loma, CA	Cape Mendocino, CA	4–4	5–6
Cape Mendocino, CA	Siuslaw River, Ore.	4–5	6–7
Siuslaw River, Ore.	Columbia, Wash.	5–6	7–8
Columbia River, Wash.	Port Townsend, Wash.	6–5	8–8
Port Townsend, Wash.	Puget Sound, Wash.	7–11	10–15

Source: Corps of Engineers (1953).

DOI: 10.1201/9781003454212-6

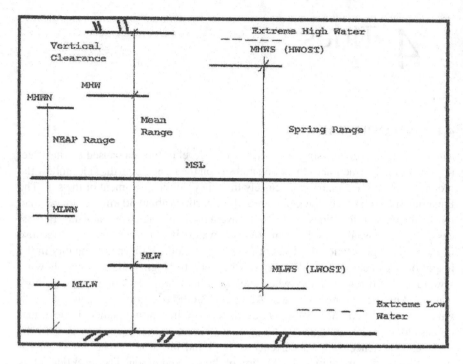

FIGURE 4.1   Summary of tidal datum.

Notes: MLW, mean low water; MLLW, mean lower low water; MSL, mean sea level; MHW, mean high water; MLWN, mean low water neap; MHWN, mean low water neap; MHWS, mean high water spring; MLWS, mean low water spring.

The design of marine structures requires the determination of a maximum water level and sometimes a minimum water level that will be experienced over the projected life of the project. These expected maximum and minimal water levels are termed the design water levels (DWLs).

The philosophy behind the requirement for a maximum DWL is that deck elevations must be above the maximum water elevation plus the crest elevation of the highest wave expected at the site. In addition, for structures such as offshore platforms it is usual practice to provide an air gap between the elevation of the crest of the highest wave and the underside of the platform. In contrast, a minimum DWL is sometimes required in the design of wharf or pile structures. In a wharf structure a decrease in the water elevation on one side will cause an increase in the lateral load that must be resisted. In a pile structure a decrease in the water level may expose the pile/soil contact to scour.

The total design water level (DWL) is the summation of the various vector components as shown in Equation 4.1.

$$DWL = d + As + Ws + Ps + Ww \tag{4.1}$$

where d is the nominal water reference depth (typically MLW); As is the astronomical tide at the time of the surge; Ws is the wind setup; Ps is the pressure or barometric setup; Ww is the wave setup.

4.2 ASTRONOMICAL TIDE-PRODUCING FORCES

The force of gravitational attraction at the surface of the earth acts in a direction toward its center-of-mass. This force also holds the ocean water confined to the earth's surface. In addition, the gravitational forces exerted by both the moon and sun also act upon the earth's ocean waters. These external forces act as tide-producing, or the so-called "**tractive**" forces. These effects are superimposed upon the earth's gravitational force and act to move the ocean waters to positions on the earth's surface directly beneath the respective celestial bodies.

High tides in the ocean are produced by the mounding action resulting from the horizontal flow of water toward two regions of the earth representing positions of maximum gravitational attraction. This maximum attraction force is produced by the combined lunar (i.e., moon)and solar (i.e., sun)gravitational forces. In contrast, low tides are created by the withdrawal of water from regions around the earth midway between the two areas of high attraction. The oscillation of seawater between high and low tides is caused by the daily (or diurnal) rotation of the earth with respect to these two tidal mounds and two tidal depressions.

4.3 ORIGIN OF THE TIDE-PRODUCING FORCES

The tidal forces at the earth's surface are the result of these two basic forces: (1) the force of gravitation exerted by the moon (and sun) upon the earth; and (2) centrifugal forces (F_c). This later force is produced by the rotation of the earth and moon (and earth and sun) around their common center-of-gravity (mass) or barycenter, refer to Figure 4.2.

The two bodies whether earth/moon or earth/sun are attracted to each other by gravitational attraction. The gravitational attraction can be determined by Newton's law of universal gravitation equation as given in the following:

$$F = \frac{GM_e M_m}{r^2} \tag{4.2}$$

where G is the gravitational constant; M_e is the mass of earth; M_m is the mass of moon; r is the distance between their respective center-of-masses.

In contrast, the two body combinations are kept apart by an equal and opposite centrifugal force. The centrifugal force trying to separate them is given in the following:

$$F_c = M_e \omega^2 l_e \tag{4.3}$$

where M_e is the mass of earth; ω is the angular rotation; l_e is the distance from earth's center-of-mass to center-of-mass of earth-moon system.

The centrifugal force is produced by their individual revolutions around their respective center-of-mass.

Assume a two-body system in which the first body represents the earth as a sphere spinning about its own axis and, in the absence of tide-producing forces, covered with a uniform layer of water. The displacements of water surface which vary with longitude will appear as tidal variation to an observer on earth moving along a constant latitude line. The second body, the moon, causes the water layer to deform into an equilibrium shape. The rotation of the earth about its own axis does not result in displacements which vary with longitude and that the only forces requiring consideration are those resulting from the revolving of the earth and moon about a common axis.

In a simple two-body case, the distance from the center of the earth to the barycenter, l_e, is given by:

$$l_e = \frac{l}{1 + \dfrac{m_e}{m_m}} \tag{4.4}$$

where l_e is the distance from body 1 to the barycenter; l is the distance between the centers of the two bodies; m_e and m_m are the masses of the earth and moon, respectively.

The effect of these forces acting on both the earth-moon and earth-sun systems is discussed in the following.

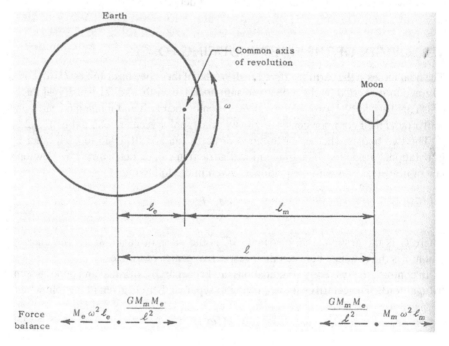

FIGURE 4.2   Earth-moon system, force balance.

(From Dean, 1966)

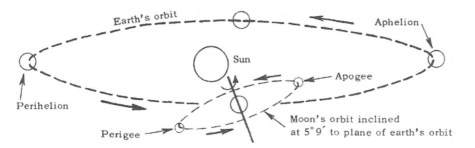

FIGURE 4.3 Motion of the earth-moon-sun system.

The centrifugal forces (F_i) tending to force the earth and moon apart are balanced by their mutual attraction forces due to gravity. The resulting net force on each of the bodies must be zero as shown in Figure 4.2. Similar motions of the earth-moon-sun system include both the revolution of the earth about the sun and the revolution of the moon about the earth. These orbits are approximately elliptical in form.

In addition to the above motions, the moon and earth each rotate about their own axes. The plane in which the earth revolves about the sun is called the ecliptic plane. The axes of rotation of the earth are inclined at 66.5 degree to the ecliptic plane. In addition, the moon's orbit about the earth is inclined at 5 degree 9 minutes to the ecliptic plane (Figure 4.3).

The positions of the moon when it is nearest to and farthest from the earth are called the perigee and apogee, respectively. At perigee the moon has its greatest tide-producing effect on the earth. In contrast, the nearest and farthest positions of the earth from the sun are called the perihelion and aphelion, respectively. A schematic illustration showing these relationships is shown in Figure 4.4.

4.4 DIURNAL AND SEMIDIURNAL TIDES

The tide prediction problem is difficult because of the complex geometry of the oceans, shorelines, and their basins. Semi-enclosed or enclosed basins tend to oscillate at some natural frequency or harmonic, so that when its natural period is close to the tide-producing period, the local tide will be amplified. Tides propagate everywhere at sea at the long-wave speed because the length of the tide wave is very long everywhere compared to the shallow ocean basins. At sea and at oceanic islands, the tidal amplitude is very small (approximately 1 foot or less). It is only in relatively shallow water such as found over the continental shelves that the amplitude begins to build. The definitions of semidiurnal (every 12 hours), mixed, and diurnal (daily) tide plots are illustrated in Figure 4.8. Tidal theories are treated in detail in a number of physical oceanography texts and papers such as Dean (1966).

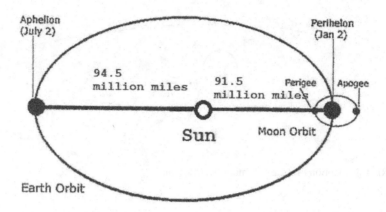

FIGURE 4.4   Interacting effects of the sun and the moon on the earth's tides.

(Adapted from "Our Restless Tides." Courtesy of the National Oceanographic and Atmospheric Administration.)

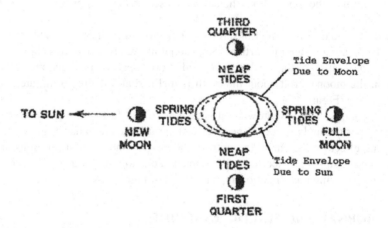

FIGURE 4.5   Sun and moon at spring and neap tides.

(Adapted from "Our Restless Tides." Courtesy of the National Oceanographic and Atmospheric Administration)

Increased tidal forces are produced when the Moon is at a position of perigee. This is the moon's closest approach to the Earth (once each month) or the Earth is at perihelion, its closest approach to the Sun (once each year).

The gravitational attractions, F_g, (and resultant tidal force envelopes) produced by the Moon and Sun reinforce each other at times of new and full moon (Figure 4.5). This increases the size of the tides, and counteract each other at the first and third quarters to reduce the tidal range. This is shown schematically in Figure 4.6. A north/south cross-section through the earth's center is shown in Figure 4.7. Looking down on the north pole of the Earth's figure (central solid circle) the two solid ellipses represent the tidal force envelope produced by the Moon in the positions of new or full moon and first and third quarters, respectively. The dashed ellipse shows the smaller tidal force envelope produced by the Sun (Figure 4.7).

4.5 DIFFERENTIAL TIDE-PRODUCING FORCES

The two forces, gravitational and centrifugal, acting on both the earth and moon always remain in balance (i.e., equal and opposite). As a result, the moon revolves in a closed orbit around the earth. This occurs without the moon either escaping from, or falling into the earth. The earth in turn does not collide with the moon. However, at specific points on, the earth, these two forces are not in equilibrium, and oceanic tides are the result.

The center of revolution of this motion of the earth and moon around their common center-of-mass (i.e., barycenter) or common axis of revolution lies at a point beneath the earth's surface, refer to Figure 4.2. This point is located on the side toward the moon, and along a line connecting the individual centers-of-mass of the earth and moon. This revolution is the two-body earth and moon system that is responsible for the centrifugal force component (F_c) necessary to the production of the tides.

4.5.1 THE EFFECT OF CENTRIFUGAL FORCES

The moon's orbital motion is responsible for one of the two force components creating the tides. As the earth and moon revolve around their common center-of-mass, the centrifugal force (F_c) produced is directed away from both the center of revolution and the moon. All points in or on the surface of the earth are affected by this component of centrifugal force.

4.5.2 THE EFFECT OF GRAVITATIONAL FORCES

The effect on earth of an external gravitational force produced by another astronomical body changes because the magnitude exerted force varies with the distance of the attracting body. According to Newton's Universal Law of Gravity, gravitational force varies inversely as the second power of the distance from the attracting body. Thus a variable influence is introduced based upon the different distances of various positions on the earth's surface from the moon's center-of-mass. The relative gravitational attraction (F_g) exerted by the moon at various positions on the earth is indicated by the length of force vectors in Figure 4.6.

4.5.3 THE NET OR DIFFERENTIAL TIDE-RAISING
FORCES: DIRECT AND OPPOSITE TIDES

The centrifugal force results from the revolution of the center-of-mass of the earth around the center-of-mass (i.e., barycenter) of the earth-moon system. Since the individual centers-of-mass of the earth and moon remain in equilibrium at constant

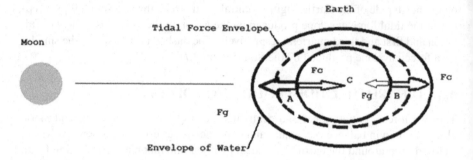

FIGURE 4.6   The combination of forces of lunar origin producing the tides.

(Adapted from "Our Restless Tides." Courtesy of the National Oceanographic and Atmospheric Administration)

distances from the barycenter. In addition, the centrifugal force acting upon the center of the earth (C) must be equal and opposite to the gravitational force exerted by the moon. The net result of the tide-producing force (F_t) at the earth's center is zero, refer to Figure 4.6.

A review of Figure 4.6 shows that point A is approximately 4000 miles nearer to the moon than is point C. The force therefore due to the moon's gravitational pull at point A is larger than at C. In addition, since the centrifugal force at A is equal to that at C, but the gravitational force at A is also larger than the centrifugal force there. The net tide-producing force at A is the difference between the gravitational and centrifugal forces and is in favor of the gravitational component—or outward toward the moon.

Point B in Figure 4.6 is approximately 4000 miles farther away from the earth's center, point C. Therefore, the moon's gravitational force at point B is less than at point C. At point C, the centrifugal force is in balance with a gravitational force which is greater than at B. The centrifugal force at B is the same as that at C. Since gravitational force is less at B than at C, it follows that the centrifugal force exerted at B must be greater than the gravitational force exerted by the moon at B. The resultant tide-producing force at this point is, therefore, directed away from the earth's center and opposite to the position of the moon.

4.5.4   THE HORIZONTAL TRACTIVE FORCE

The effect of the moon's small gravitational attraction is superimposed upon the larger force of earth's gravity. The tidal force produced by the moon's gravitational attraction at any point on the earth's surface, can be resolved into two components of force. One component is perpendicular to the earth's surface. The other component is tangent to the earth's surface. The tangential component of force is the actual mechanism for producing tides. This tidal force is zero at the points on the earth's surface directly beneath and on the opposite side of the earth from the moon. Any water accumulated in these locations by this tangential force causes flow from other points on the earth's surface tends to remain in a stable configuration, or tidal "mound." As a result, the tides are produced by the horizontal component of the force of the moon which acts to move the waters over the earth's surface toward the sublunar and opposite points.

4.5.5   THE TIDAL FORCE ENVELOPE

An ellipsoid figure would be formed if the ocean waters were completely able to respond to the directions and magnitudes of the tractive forces at various points on the surface of the earth. The longest axis of the ellipsoid extends both toward and directly away from the moon, and the shortest axis is centered, and at right angle to, the major axis. The two tidal humps and two tidal depressions are represented in this force envelope by the directions of the major axis and minor axis of the ellipsoid, respectively. The daily rotation of the solid earth with respect to these two tidal humps and two depressions may be conceived to be the cause of the tides.

As the earth rotates once in each 24 hours, it would be expected to find a high tide followed by a low tide at the same place 6 hours later; then a second high tide after 12 hours, a second low tide 18 hours later, and finally a return to high water at the expiration of 24 hours. This would be the case if the following were true: (1) a smooth, continent-free earth were covered to a uniform depth with water, (2) the tidal envelope of the moon alone were being considered, (3) the positions of the moon and sun were fixed and invariable in distance and relative orientation with respect to the earth, and (4) there were no other accelerating or retarding influences affecting the motions of the waters of the earth. These conditions are far from the situation that exists.

First, the tidal force envelope produced by the moon's gravitational attraction is accompanied by a tidal force envelope of considerably smaller amplitude produced by the sun (Figure 4.6). The tidal force exerted by the sun is a composite of the sun's gravitational attraction and a centrifugal force component created by the revolution of the earth's center-of-mass around the center-of-mass of the earth-sun system, in an comparable manner to the earth-moon relationship. The position of this force envelope shifts with the relative orbital position of the earth in respect to the sun. Because of the great differences between the average distances of the moon (238,855 miles) and sun (92,900,000 miles) from the earth, the tide-producing force of the moon is approximately 2.5 times that of the sun.

Second, there exists a wide range of astronomical factors in the production of the tides. Some of these factors are the following: (1) the changing distances of the moon from the earth, (2) the changing distances of the earth from the sun, (3) the angle which the moon in its orbit makes with the earth's equator, (4) the superposition of the sun's tidal envelope of forces upon that caused by the moon, and (5) the variable phase relationships of the moon. Some of the principal types of tides resulting from these purely astronomical influences are describe below.

4.6   TIDAL CURRENTS

Tidal current relates to the periodic horizontal movement of the ocean water, both near the coast and offshore. The tidal current is distinct from the continuous, stream-flow type of ocean current. Tides are important because it indicates when the level of seawater is located at the base of a bluff leading to erosion by waves. The bluff retreat rate is directly tied to the wave energy and the type of material comprising the shoreline.

4.7   VARIATIONS IN THE RANGE OF THE TIDES

The difference in the height between consecutive high and low tides occurring at a given location is known as the range. The range of the tides at any location is subject to many variable factors. Those influences of astronomical origin will first be described.

4.7.1   LUNAR PHASE EFFECT: SPRING AND NEAP TIDES

The gravitational forces of both the moon and sun act upon the waters of the earth. The gravitational attraction of moon and sun either acts along a common line or at changing angles relative to each other. This occurs because the moon's position changes with respect to the earth and sun during its monthly cycle of phases (29.53 days) (Figure 4.3).

The gravitational attractions of the moon and sun act to reinforce each other when the moon is either at a new phase or full phase. The observed high tides at this time are higher and low tides are lower than average because the resultant or combined tidal force has also increased. Such greater-than-average tides resulting at the syzygy positions (i.e., a conjunction or opposition, especially of the moon with the sun) of the moon are known as spring tides (Figure 4.5).

The gravitational attractions of the moon and sun upon the waters of the earth are exerted at first- and third-quarter phases of the moon and at right angles to each other. Each force tends in part to counteract the other. In the tidal force envelope representing these combined forces, both maximum and minimum forces are reduced. High tides are lower and low tides are higher than average. Such tides of diminished range are called neap tides.

4.7.2   PARALLAX EFFECTS (MOON AND SUN)

Since the moon follows an elliptical path (Figure 4.4), the distance between the earth and moon will vary throughout the month by about 31,000 miles. The moon's tide-producing force acting on the earth's waters will change in inverse proportion to the third power of the distance between the earth and moon. This is in accordance with the previously mentioned variation of Newton's Law of Gravitation. Once each month, when the moon is closest to the earth (perigee), the tide-generating forces will be higher than usual, thus producing above-average ranges in the tides. Approximately two weeks later, when the moon (at apogee) is farthest from the earth, the lunar tide-raising force will be smaller, and the tidal ranges will be less than average. Similarly, in the earth-sun system, when the earth is closest to the sun (perihelion), about January 2 of each year, the tidal ranges will be enhanced, and when the earth is farthest from the sun (aphelion), around July 2, the tidal ranges will be reduced.

Tidal ranges increase when perigee, perihelion, and either the new or full moon occur at approximately the same time. In contrast, when apogee, aphelion, or the first- or third-quarter of the moon coincide at approximately the same time, reduced tidal ranges will normally occur.

4.7.3 LUNAR DECLINATION EFFECTS: DIURNAL INEQUALITY

The plane of the moon's orbit is inclined approximately 5 degrees to the plane of the earth's orbit (i.e., the ecliptic). This results in the moon monthly revolution around the earth remaining very close to the ecliptic. In turn the ecliptic is inclined 23.5 degrees to the earth's equator, north and south of which the sun moves once each half year to produce the seasons. In similar fashion, the moon, in making a revolution around the earth once each month, passes from a position of maximum angular distance north of the equator to a position of maximum angular distance south of the equator during each half month. (Angular distance perpendicularly north and south of the celestial equator is termed declination.) Thus twice each month, the moon crosses the equator. In Figure 4.7, this condition is shown by the dashed position of the moon. The corresponding tidal force envelope due to the moon is depicted, in profile, by the dashed ellipse.

Since the points A and A' lie along the major axis of this ellipse, the height of the high tide represented at A is the same as that which occurs as this point rotates to position A' some 12 hours later (Figure 4.7). When the moon is over the equator—or

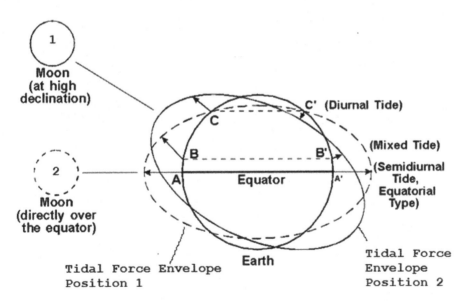

FIGURE 4.7 The Moon's declination effect (change in angle with respect to the equator) and the diurnal inequality; semidiurnal, mixed, and diurnal tides.

(Adapted from "Our Restless Tides." Courtesy of the National Oceanographic and Atmospheric Administration)

at certain other force-equalizing declinations—the two high tides and two low tides on a given day are at similar height at any location. Successive high and low tides are then also nearly equally spaced in time, and occur twice daily (see top diagram in Figure 4.8a). This is known as semidiurnal type of tides.

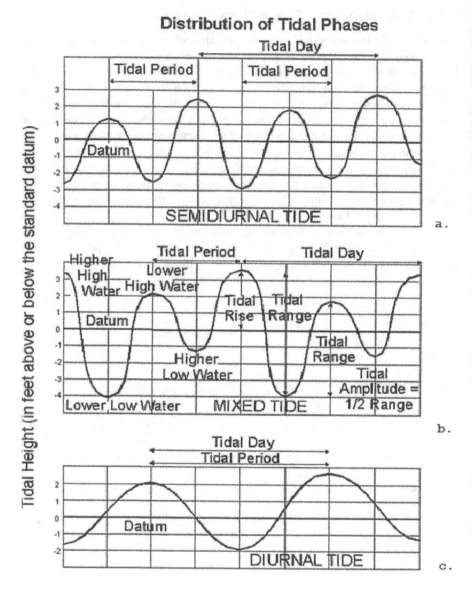

FIGURE 4.8 The Moon's declinational effect in production of semidiurnal, mixed, and diurnal tides.

(Adapted from "Our Restless Tides." Courtesy of the National Oceanographic and Atmospheric Administration)

However, with the changing angular distance of the moon above or below the equator (represented by the position of the small solid circle in Figure 4.7) the tidal force envelope produced by the moon is canted, and difference between the heights of two daily tides of the same phase begin to occur. Variations in the heights of the tides resulting from the changes in the declination angle of the moon and in the corresponding lines of gravitational force action give rise to a phenomenon known as the diurnal inequality.

In Figure 4.7, point B is beneath a bulge in the tidal force envelope. One-half day later, at point B' it is again beneath the bulge, but the height of the tide is obviously not as great as at B. This situation gives rise to a twice-daily tide displaying unequal heights in successive high or low waters, or in both pairs of tides. This type of tide is known as a mixed tide (Figure 4.8b).

Finally, the point C is seen to lie beneath a portion of the tidal force envelope (Figure 4.7). One-half day later, this point rotates to position C'. A review of Figure 4.7 shows that it is seen to lie above the force envelope. At this location, therefore, the tidal forces present produce only one high water and one low water each day. The resultant diurnal type of tide is shown in the bottom diagram of Figure 4.8c.

4.8 FACTORS INFLUENCING THE LOCAL HEIGHTS AND TIMES OF ARRIVAL OF THE TIDES

4.8.1 PREDICTION OF THE TIDES

In the preceding discussions of the tide-generating forces, it has been emphasized that the tides actually observed differ appreciably from the idealized, equilibrium tide. Nevertheless, because the tides are produced essentially by astronomical forces of harmonic nature, a definite relationship exists between the tide-generating forces and the observed tides.

Because of both the numerous uncertain and, unknown factors of local control, it is not feasible to predict tides purely from a knowledge of the positions and movements of the moon and sun obtained from astronomical tables. A partially empirical approach based upon actual observations of tides in many areas over an extended period of time is necessary. To achieve maximum accuracy in prediction, a series of tidal observations at one location ranging over at least a full 18.6-year tidal cycle is required. Within this period, all significant astronomical modifications of tides will occur. In the Pacific Northwest, there are two high tides and two low tides in a 25-hour period, the normal tidal range (i.e., the difference between high and low tides) is 3.3 m (11 feet) but can be as much as 5.8 m (19 feet).

4.9 SUMMARY

The force of attraction exerted by the Moon's and the Sun's on the Earth's hydrosphere causes ocean tides. The Moon's tide-generating force is about twice that of the Sun because it is much closer to the Earth. The Moon and Earth revolve around a common center-of-mass (i.e., barycenter) that is located inside the Earth.

This produces a centrifugal force that balances the forces of attraction. In the equilibrium tide model, two tidal bulges are developed because masses on the Earth's surface are acted on unequally by gravitation and centrifugal forces. One bulge faces the Moon and the other is directly opposite. As the Moon orbit makes excursions north and south of the equator as it revolves around its common center-of-mass with the Earth. When the Moon is over the equator the Earth passes through unequal successive tidal bulges. This condition produces different elevations in successive high and low tides and unequal tidal ranges. This tidal condition is called a semidiurnal inequality.

The Moon's tide-generating force can be enhanced or retarded by the Sun. When the Moon, Earth, and Sun are aligned (i.e., a position called syzygy), the Sun's effects are additive. As a result spring tides and large tide ranges occur. When the Moon, Earth, and Sun are at right angles (i.e., quadratic position) the Sun's effects diminish the Moon's tide-generating forces. This condition causes neap tides and results in relatively small tidal ranges. Mean tides and average tidal ranges occur between syzygy and the quadratic positions. Due to the elliptical orbits of the Moon and Earth, the height and range of the tides increase when the Moon is proximate to the Earth (perigean tides) and the Earth is close to the Sun (perihelion tides).

REFERENCES

Corps of Engineers. (1953). Shore Protection Planning and Design, The Bulletin of the Beach Erosion Board, Special Issue No. 2, U.S. Army, March: 360 pp.

Dean, R. G. (1966). Tides and harmonic analysis. In Ippen, A. T. (Ed.), *Estuary and Coastline Hydrodynamics*. New York: McGraw-Hill, pp. 197–230.

Dean, R. G. Sea Level Rise Viewer. Crescent City Tide Gauge. Accessed 1/19/21. https://coast.noaa.gov/slr/#/layer/slr/0/-13795063.467339959/5125122.28856575/10/satellite/75/0.8/2050/interHigh/midAccretion.

National Oceanic and Atmospheric Administration (NOAA). (2022). Our Restless Tides. *NOAA Tides and Currents*. Silver Spring, MD: NOAA. http://tidesandcurrents.noaa.gov/restles1.html

5 Ocean Waves

5.1 INTRODUCTION

The basic hydrodynamic equations governing wave motion are momentum and continuity equations for an incompressible irrotational inviscid fluid with constant density along with the appropriate boundary conditions. A schematic figure presenting the coordinate system and velocity components for an idealized case is presented in Figure 5.1.

The momentum equations can be combined and integrated to give the Bernoulli equation for unsteady motion. This relationship is given by the following:

$$\frac{\partial \phi}{\partial t} + \frac{1}{2}\left(u^2 + v^2 + w^2\right) + \frac{P}{\rho} + gy = f(t), \tag{5.1}$$

where $f(t)$ is an arbitrary equation of time which can be set to zero. Assuming all motions are small (i.e., small-amplitude wave theory) and restricting the solution to two dimensions results in the following.

$$-\frac{\partial \phi}{\partial z} + \frac{P}{\rho} + gz = 0. \tag{5.2}$$

This is the general equation of motion applied to the development of the small-amplitude theory of water waves. Assume an idealized wave as shown in Figure 5.2 is moving to the right (+x direction).

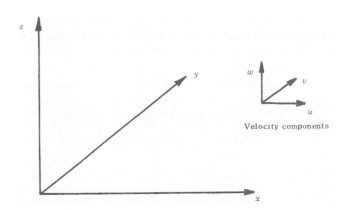

FIGURE 5.1 Coordinate system and velocity components.

DOI: 10.1201/9781003454212-7

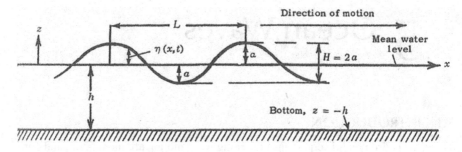

FIGURE 5.2 Small-amplitude wave system.

where:

> h—distance from mean water level to bottom
> $\eta(x,t)$—instantaneous vertical displacement of the water surface above mean
>   water level
> a—wave amplitude
> H—wave height $= 2a$ for small amplitude waves
> L—wavelength
> T—wave period
> C—velocity of wave propagation (phase velocity) $= L/T$
> k—wave number $= {2\pi}/{L}$
> σ—wave angular frequency $= {2\pi}/{T}$

The differential to be satisfied is in the region $-h \le z \le \eta$ and $-\infty < x < +\infty$ and is described by the Laplace equation as shown below for two dimensions.

$$\frac{\partial^{3}\phi}{\partial x^{2}} + \frac{\partial^{2}\phi}{\partial z^{2}} = 0, \tag{5.3}$$

where ϕ is a scalar function $\phi(x, z, t)$.

To solve this equation it is necessary to determine the boundary conditions for the present case. In the present case, the upper boundary is not only moving but its position is not specified and depends on the solution of the problem. In contrast, if the bottom is a fixed impermeable horizontal boundary then the boundary condition is as follows:

$$\omega = -\frac{\partial\phi}{\partial z} = 0 \text{ at } z = -h. \tag{5.4}$$

In contrast, the boundary condition on the surface is the following:

$$\eta = \frac{1}{g}\left[\frac{\partial\eta}{\partial t}\right]_{z=\eta} \text{ on } z = 0. \tag{5.5}$$

The corresponding velocity potential for a progressive wave traveling in the negative x direction is given as follows (Eagleson and Dean, 1966).

$$\phi = -\frac{ag\cosh k(h+z)}{\sigma \cosh kh}\sin(kx - \sigma t). \qquad (5.6)$$

The above system of equations represents the general free-surface-wave problem. Other boundary conditions, such as vertical walls and piles, impose further conditions on the solutions.

5.2 WAVE CHARACTERISTICS

Ocean surface waves refer to a moving succession of irregular crests and troughs on the ocean surface. A slight disturbance of the sea surface will be exhibited as tiny round ripples or capillary waves as the wind begins to blow (Figure 5.3a). Capillary waves have very small wavelengths (L), short periods (T), and small wave heights (H). To return the sea surface to an undisturbed glassy state requires damping these capillary waves. The damping force is surface tension. This damping or restoring force results in the propagation of the wave in a horizontal direction. This is similar to a small wave traveling along a tightly stretched piece of fabric.

Waves grow in length, height, and energy as the wind continues to blow. The growth of waves results because the disturbed sea surface is increasingly exposed to the wind. When the wavelengths exceed 1.74 cm gravity replaces surface tension as the dominant restoring force. The resultant wave form is called a gravity wave (Figure 5.3b). A short choppy sea characteristically then develops because of the interactions between different waves creating a variety of wave forms with different wavelengths. Wave height increases more rapidly than wavelength as more energy is imparted to the waves by the wind. This process continues until the resultant wave

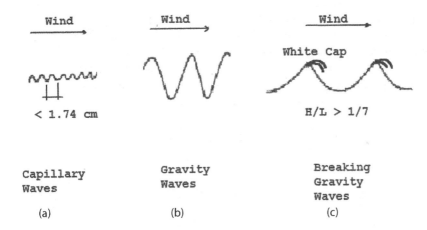

FIGURE 5.3 Development of waves on an initially smooth sea surface in response to a sustained unidirectional wind. (a) Capillary waves, (b) gravity waves, and (c) breaking gravity waves.

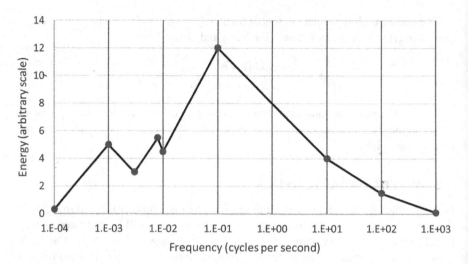

FIGURE 5.4 Classification of surface waves and schematic representation of their power spectrum.

(Adapted from US Army, Corps of Engineers, 1977; Courtesy of the USA COE)

steepness causes them to break (Figure 5.3c). In breaking waves, or whitecaps, the energy received from the wind is balanced by the energy lost in breaking.

The same fundamental laws of physics controlling sound and light waves also apply to ocean waves. Waves represent the propagation of mechanical energy due to wind, earthquakes, volcanic activity, landslides, meteorological phenomena, or even by other waves. An important property of waves under ideal conditions is that there is no net displacement of the particles set in motion by a wave. Water molecules affected by an ocean wave describe a circular orbit but undergo only a slight net forward displacement. The orbital motion of particles is characteristic of waves occurring between fluids of different densities. Waves move along the air/water interface but also along any pycnocline. There is a complete spectrum of waves ranging from small capillary waves (2 cm long) to the tides with wavelengths of thousands of kilometers. The properties of ocean surface waves and a schematic representation of their power spectrum are presented in Figure 5.4.

5.3   GRAVITY WAVE GENERATION

5.3.1   Introduction

The surface-wave data can be used to calculate the water velocity, acceleration, and pressure beneath the waves for engineering purposes, three wave theories have been developed to accomplish this. These theories are the following: (1) Linear (Airy) wave theory, and two nonlinear methods, (2) Stokes wave theory, and (3) Cnoidal wave theory. Only the linear wave theory will be used in this chapter.

5.3.2 LINEAR WAVE THEORY (I.E., AIRY THEORY)

A simple theory of wave motion, known as the Airy wave theory, was developed by G.B. Airy in 1842. This theory assumes a sinusoidal wave form whose height H is small in comparison with the wavelength Δ and the water depth h (Figure 5.2). The theory is useful for preliminary calculations and for illustrating the basic characteristics of wave-induced water motion. It also serves as a basis for the statistical representation of waves and the induced water motion during storm conditions.

A surface wave amplitude (η) of point A as shown in Figure 5.2 as a function of time can be described by the following equation.

$$\eta = \frac{H}{2}\cos(kx - \omega t + \psi). \tag{5.7}$$

5.3.2.1 Fluid Particle Velocity

The horizontal velocity (u) and vertical velocity (v) of a water particle at point A (x,y) and time t are expressible by relationships that have been developed by a number of authors including Kinsman (1965) and McCormick (1973).

$$u = \frac{wH}{2}\frac{\cosh ky}{\sinh kh}\cos(kx - wt + \Psi), \tag{5.8}$$

$$u = \frac{H}{2}\frac{gT}{L}\left[\frac{\cosh[2\pi(z+d)]\big/L}{\cosh(2\pi d\big/L)}\right]\cos\left(\frac{2\pi x}{L} - \frac{2\pi t}{T}\right), \tag{5.9}$$

$$v = \frac{wH}{2}\frac{\sinh ky}{\sinh kh}\sin(kx - wt + \psi), \tag{5.10}$$

$$v = \frac{H}{2}\frac{gT}{L}\left[\frac{\cosh[2\pi(z+d)]\big/L}{\cosh(2\pi d\big/L)}\right]\sin\left(\frac{2\pi x}{L} - \frac{2\pi t}{T}\right), \tag{5.11}$$

where:

k—wave number = $2\pi/\lambda$
ω—angular radian frequency = $2\pi/T$
T—wave period
λ—wavelength
H—wave amplitude
X—horizontal direction
Ψ—Phase angle

The above equations give the local fluid velocity components at any height ($z+d$) above the bottom. The velocities are harmonic (i.e., integer multiple of the fundamental frequency) in both x and t. For a given value of the phase angle $\theta = (kx - wt)$, the hyperbolic functions cosh and sinh, in z result in an approximate exponential decay of the velocity components magnitude with increasing distance below the free surface. The maximum positive horizontal velocity occurs when $\theta = 0, 2\pi$, etc. In contrast, the maximum horizontal velocity in the negative direction occurs when $\theta = \pi, 3\pi$, etc. The maximum positive vertical velocity occurs when $\theta = \pi/2, 5\pi/2$, etc. The maximum vertical velocity in the negative direction occurs when $\theta = 3\pi/2, 7\pi/2$, etc.

5.3.2.2 Fluid Particle Acceleration

The local fluid particle accelerations are obtained by differentiating Equations (5.8) and (5.10) with respect to t as presented below:

$$a_z = -\frac{g\pi H}{L}\left[\frac{\sinh\left(\frac{2\pi(z+d)}{L}\right)}{\cosh\left(\frac{2\pi d}{L}\right)}\right]\cos\left(\frac{2\pi x}{L} - \frac{2\pi t}{T}\right), \tag{5.12}$$

$$a_x = -\frac{g\pi H}{L}\left[\frac{\cosh\left(\frac{2\pi(z+d)}{L}\right)}{\cosh\left(\frac{2\pi d}{L}\right)}\right]\sin\left(\frac{2\pi x}{L} - \frac{2\pi t}{T}\right). \tag{5.13}$$

Positive and negative values of the horizontal and vertical fluid accelerations for various values of $\theta = \dfrac{2\pi}{L} - \dfrac{2\pi t}{T}$ are also shown in Figure 5.5.

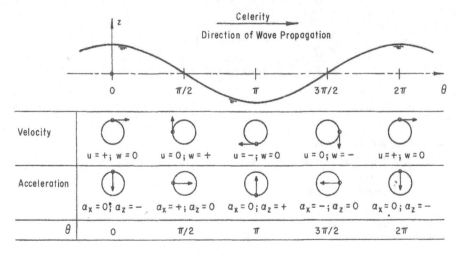

FIGURE 5.5   Local field velocities and accelerations.

(Courtesy of U.S. Army Corps of Engineers, Shoreline Protection Manual)

5.3.2.3 Water Particle Displacements

The displacement of individual water particles within a wave is an aspect of linear wave mechanics (Figure 5.5). Water particles typically move in elliptical paths in shallow or transitional water and in circular paths in deep water (Figure 5.6). If the particle position is at the center of the ellipse or circle, then the vertical particle displacement cannot exceed one-half the wave height. Therefore since the wave height is assumed to be small, the displacement of any fluid particle is small. Integration of Equations (5.12) and (5.13) gives the horizontal and vertical particle displacements from the mean position, respectively.

$$\xi = -\frac{gT^2H}{4\pi L}\left[\frac{\cosh\left(2\pi(z+d)\big/L\right)}{\cosh\left(2\pi d\big/L\right)}\right]\sin\left(\frac{2\pi x}{L}-\frac{2\pi t}{T}\right), \qquad (5.14)$$

$$\varsigma = -\frac{gT^2H}{4\pi L}\left[\frac{\sinh\left(2\pi(z+d)\big/L\right)}{\cosh\left(2\pi d\big/L\right)}\right]\cos\left(\frac{2\pi x}{L}-\frac{2\pi t}{T}\right). \qquad (5.15)$$

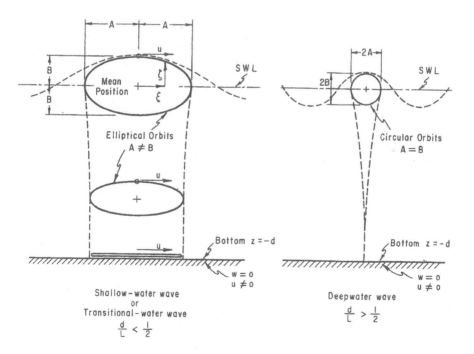

FIGURE 5.6 Water particle displacements from mean position for shallow water and deep water waves.

(Courtesy of U.S. Army Corps of Engineers, Shoreline Protection Manual)

The above equations can be simplified by substituting Equation (5.16) into Equations (5.14) and (5.15).

$$\left(\frac{2\pi}{T}\right)^2 = \frac{2\pi g}{L}\tanh\left(\frac{2\pi d}{L}\right), \tag{5.16}$$

$$\xi = -\frac{H}{2}\left[\frac{\cosh\left(2\pi(z+d)\big/L\right)}{\sinh\left(2\pi d\big/L\right)}\right]\sin\left(\frac{2\pi x}{L} - \frac{2\pi t}{T}\right), \tag{5.17}$$

$$\varsigma = +\frac{H}{2}\left[\frac{\sinh\left(2\pi(z+d)\big/L\right)}{\sinh\left(2\pi d\big/L\right)}\right]\cos\left(\frac{2\pi x}{L} - \frac{2\pi t}{T}\right). \tag{5.18}$$

Rearranging the above equations

$$\sin^2\left(\frac{2\pi x}{L} - \frac{2\pi t}{T}\right) = \left[\frac{\xi}{a}\frac{\sinh\left(2\pi d\big/L\right)}{\cosh\left(2\pi(z+d)\big/L\right)}\right]^2, \tag{5.19}$$

$$\cos^2\left(\frac{2\pi x}{L} - \frac{2\pi t}{T}\right) = \left[\frac{\varsigma}{a}\frac{\sinh\left(2\pi d\big/L\right)}{\sinh\left(2\pi(z+d)\big/L\right)}\right]^2, \tag{5.20}$$

and adding together gives the following:

$$\frac{\xi^2}{A^2} + \frac{\varsigma^2}{B^2} = 1, \tag{5.21}$$

$$A = +\frac{H}{2}\left[\frac{\cosh\left(2\pi(z+d)\big/L\right)}{\sinh\left(2\pi d\big/L\right)}\right], \tag{5.22}$$

$$B = +\frac{H}{2}\left[\frac{\sinh\left(2\pi(z+d)\big/L\right)}{\sinh\left(2\pi d\big/L\right)}\right]. \tag{5.23}$$

Equation (5.21) is an equation of an ellipse with a major horizontal semi-axis equals to A and a minor vertical semi-axis equals to B. The lengths of A and B

are measures of the horizontal and vertical displacements of the water particles. Therefore, the water particles by linear wave theory move in closed orbits.

A review of Equations (5.22) and (5.23) shows that A and B are equal and particle paths are circular for deep water conditions. The equations become the following:

$$A = B = \frac{H}{2} e^{2\pi z/L} \text{ for } d/L > 1/2. \tag{5.24}$$

For shallow-water conditions, the equations become

$$A = \frac{H}{2} \frac{L}{2\pi d} \text{ for } d/L < 1/25, \tag{5.25}$$

$$B = \frac{H}{2} \frac{z+d}{d}. \tag{5.26}$$

The water particle in deep water moves in circular orbits. The shallower the water, the flatter the ellipse. The water particle displacement amplitude decreases exponentially with depth. In deep water regions, the particle displacement become small relative to the wave height at a depth equal to one-half the wavelength below the free surface; that is, when $z = L_o/2$ as shown in Figure 5.6. For shallow regions, horizontal particle displacement near the bottom can be large (Figure 5.6). This is apparent in offshore regions seaward of the breaker zone where wave action and turbulence lift bottom sediments into suspension.

These terms are related to each other by the following equation which expresses the circular frequency of the wave as shown below. The relationship between radian frequency and wave number depends on water depth and is given in Equation (5.27).

$$\omega = \left[gk \ tanhkh \right]^{1/2}, \tag{5.27}$$

where:

g—acceleration due to gravity
h—water depth.

This relationship is called the wave dispersion equation. For large values of kh, tanh kh goes to 1 as shown in Figure 5.7 then Equation (5.27) becomes

$$\omega^2 = gk. \tag{5.28}$$

If the term $(kx - \omega t)$ in Equations (5.8) and (5.10) remains unchanged at time $(t + \Delta t)$ as shown in Figure 5.8, then the wave amplitude (H) remains the same.

This can occur if Δx equals the following:

$$\Delta x = \left(\frac{\omega}{k} \right) \Delta t, \tag{5.29}$$

then

$$kx - \omega t = k \left(x + \Delta x \right) - \omega \left(t + \Delta t \right). \tag{5.30}$$

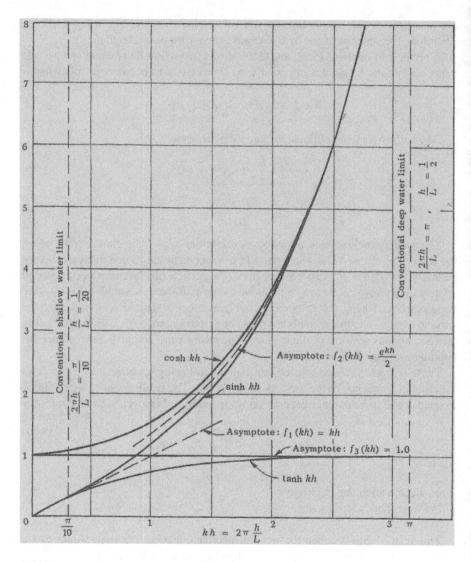

FIGURE 5.7   Hyperbolic functions and asymptotes.

The surface wave described by Equation (5.7) can therefore easily be seen to represent a fixed wave form propagating to the right with a speed (i.e., phase velocity) or celerity c. The celerity (c) can be related to wavelength (L) and period (T) using Equation (5.31).

$$c = \frac{L}{T} = \frac{w}{k}.$$
(5.31)

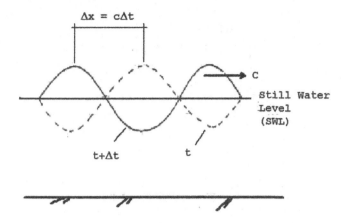

FIGURE 5.8 Propagating waves.

Substituting Equation (5.27) into Equation (5.31) gives the following relationship.

$$c = \left(\frac{g}{k} \tan h\, kh\right)^{1/2} = \left[\frac{gL}{2\pi} \tan h\left(\frac{2\pi d}{L}\right)\right]^{1/2}. \tag{5.32}$$

From Equation (5.30), Equation (5.31) can be rewritten as follows.

$$c = \frac{gT}{2\pi} \tanh\left(\frac{2\pi d}{L}\right) = \frac{gT}{kL} \tanh(kd). \tag{5.33}$$

An expression for wavelength (L) as a function of depth (d) and wave period (T) can be obtained:

$$L = \frac{gT^2}{2\pi} \tanh \frac{2\pi d}{L}. \tag{5.34}$$

Equation (5.34) has L on both sides of the equation. To solve this equation two methods may be employed. The first method involves an iterative procedure where a value of L is guessed and then calculated using Equation (5.34). The average between the guessed L and the calculated L is then used. This process is repeated until closure is achieved. The second procedure is an approximate relationship presented by Eckart (1952) to solve Equation (5.34). This approximate expression is presented in Equation (5.35).

$$L = \frac{gT^2}{2} \left[\tan h\left(\frac{4\pi^2}{T^2} \frac{h}{g}\right)\right]^{1/2}. \tag{5.35}$$

5.3.2.4 Wave Classification
Using Equation (5.34) harmonic (i.e., gravity) waves can be classified by the water depth in which they travel (Table 5.1). These waves are shallow, intermediate, and

TABLE 5.1

Water Depth Classification

d/L_o	$\tanh(2\pi d/L_o)$
0.1	0.556
0.2	0.849
0.5	0.996
1.0	0.999

deep water waves depending on the ratio of water depth to wavelength (d/L). This ratio also leads to three-phase velocity classifications. The following classifications are made based on the magnitude of (d/L) and the resulting limiting values taken by the relationship $\tanh(2\pi d/L)$.

Classification	d/L	$2\pi d/L$	$\tanh(2\pi d/L)$
Deep water	>1/2	$>\pi$	≈ 1
Transitional	1/25–1/2	$1/4$–π	$\tanh(2\pi d/L)$
Shallow water	<1/25	<1/4	$\approx 2\pi d/L$

Deep water occurs at an infinite depth (d). In reality, the relationship $\tanh(2\pi d/L)$ as shown in Figure 5.7 approaches a value of 1 at a much smaller value of d/L_o as shown in Table 5.2.

Thus when $d/L > 0.5$ the wave characteristics (c, L) are effectively independent of depth. The period (T) remains constant and independent of depth for harmonic waves.

A. For Deep Water $kh > \pi$, $1/2 < h/L < \infty$
   For this range of h/L indicated.
   $\tanh kh \sim 1$.

$$c_o = \left(\frac{g}{k}\right)^{1/2} = \left(\frac{L_o g}{2\pi}\right)^{1/2} = \frac{L_o}{T},$$
(5.36)

TABLE 5.2

Simplification for Deep Water

Units	SI	English
$C_o = \dfrac{gT}{2\pi}$	1.56 T (m/s)	5.12 T (ft/s)
$L_o = \dfrac{gT^2}{2\pi}$	1.56 T^2 (m)	5.12 T^2 (ft)

Note: $g/2\pi = 1.56$ m/s $= 5.12$ ft/s.

where:

g—acceleration due to gravity
c_o—celerity in deep water
L_o—wavelength in deep water
k = wave number = $2\pi/L$

A review of Equation (5.36) indicates that the celerity (c_o) of a wave (speed) in deep water depends upon its wavelength (L). Waves with longer wavelengths travel faster than waves with shorter wavelengths. At a site a distance from a source of wave generation in deep water long wavelength waves tend to arrive first followed by waves of shorter wavelengths. This spreading out of waves is called dispersion.

Combining Equation (5.31) with Equation (5.36) gives the following for deep water:

$$c_o = \frac{Tg}{2\pi},\tag{5.37}$$

$$\text{or } L_o = \frac{T^2 g}{2\pi}.\tag{5.38}$$

Note that combining Equations (5.37) and (5.38) shows that $c_o\, \alpha.\sqrt{L_o}$.

The speed of a swell depends upon the square root of its wavelength. Therefore in deep water, longer waves move faster than short waves as they propagate from a storm area. This process is causes dispersion and explains why the first waves to reach a distinct beach from a large storm are the longer waves.

B. For Shallow Water, $kh < \pi/10$ or $h/L < 1/20$

$$\tanh kh \sim kh = \frac{2\pi h}{L},\tag{5.39}$$

$$c = \left(\frac{2\pi gh}{kL}\right)^{1/2} = (gh)^{1/2}.\tag{5.40}$$

Thus the speed of a shallow water wave is the following:

$$c = \sqrt{gh},\tag{5.41}$$

where h is the depth.

5.3.2.5 Subsurface Pressure

The gage pressure is the difference between actual pressure and atmospheric pressure at any place (x,y) and time t. This is the result of the dynamic pressure due to the wave and from the hydrostatic (static pressure) contribution. This pressure can be determined using the following expression.

$$u' = \rho_w\, g\, \frac{H \cosh ky}{2 \cosh kh} \cos(kx - \omega t) + \rho_w\, g(h - y) + p_a,\tag{5.42}$$

or

$$u' = \rho_w g \left[\frac{\cosh\left(2\pi(z+d)\big/L\right)}{\cosh\left(2\pi d\big/L\right)} \right] \cos\left(\frac{2\pi x}{L} - \frac{2\pi t}{T}\right) - \rho g z + p_a. \qquad (5.43)$$

The pressure is the gage pressure defined as follows:

$$u = u' - p_a = \rho_w g \left[\frac{\cosh\left(2\pi(z+d)\big/L\right)}{\cosh\left(2\pi d\big/L\right)} \right] \cos\left(\frac{2\pi x}{L} - \frac{2\pi t}{T}\right) - \rho g z, \qquad (5.44)$$

where:

ρ_w = mass density of water
u' = total or absolute pressure
u = gage pressure
p_a = atmospheric pressure

The first term of Equation (5.42) represents a dynamic component due to acceleration, while the second term is the static component of pressure.

Let η be defined as follows:

$$\eta = \frac{H}{2} \cos\left(\frac{2\pi x}{L} - \frac{2\pi t}{T}\right), \qquad (5.45)$$

then

$$u = \rho \eta g \left[\frac{\cosh\left(2\pi(z+d)\big/L\right)}{\cosh\left(2\pi d\big/L\right)} \right] - \rho g z. \qquad (5.46)$$

Let the ratio equals K_z
$$K_z = \frac{\cosh\left(2\pi(z+d)\big/L\right)}{\cosh\left(2\pi d\big/L\right)}. \qquad (5.47)$$

It is termed as the pressure response factor. Therefore, Equation (5.44) can be written as follows:

$$u = \rho g\left(\eta K_z - z\right). \qquad (5.48)$$

5.3.2.6 Velocity of a Wave Group

A group of waves (i.e., wave train) travels at a speed which is normally not the same as individual waves within the group. This speed of the group is called the group velocity C_g. In contrast, the speed of individual waves is the phase velocity or wave celerity (c).

The group velocity can be modeled as the interaction of two sinusoidal wave trains moving in the same direction with slightly different wavelengths and periods. The equation of the water surface can then be given by the following:

$$\eta = \eta_1 + \eta_2 = \frac{H}{2}\cos\left(\frac{2\pi x}{L_1} - \frac{2\pi t}{T_1}\right) + \frac{H}{2}\cos\left(\frac{2\pi x}{L_2} - \frac{2\pi t}{T_2}\right), \tag{5.49}$$

where η_1 and η_2 are the contributions of each of the two components. These individual components may be summed since the linear wave theory was used. For simplicity, assume the heights of both wave components are equal. Since the wavelengths of the two components waves, L_1 and L_2 have been assumed slightly different for some values of x at a given time, the two components will be in phase and the height observed will be $2H$. For other values of x, the two waves will be completely out of phase and the resultant wave height will be zero. The surface profile made up as the sum of the two sinusoidal waves is given by Equation (5.49) and is shown in Figure 5.9. The waves shown in Figure 5.9 appear to be traveling in groups described by the equation of the envelope curves.

$$\eta_{\text{envelope}} = H\cos\left[\pi\left(\frac{L_2 - L_1}{L_1 L_2}\right)x - \pi t\left(\frac{T_2 - T_1}{T_1 T_2}\right)\right]. \tag{5.50}$$

It is the speed of these groups (i.e., the velocity of propagation of the envelope curves) that represents the group velocity. The limiting speed of the wave groups as they become large (i.e., as the wavelength L_1 approaches L_2 and consequently the

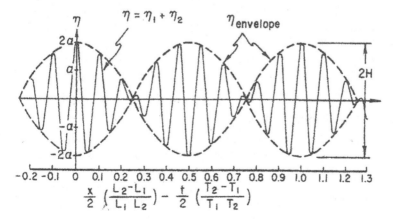

FIGURE 5.9 Formation of wave groups by the addition of two sinusoids having different periods.

(U.S. Army Corps of Engineers, Shoreline Protection Manual, 1977)

wave period T_1 approaches T_2) is the group velocity and can be shown to be equal to the following:

$$C_g = \frac{1}{2}\frac{L}{T}\left[1 + \frac{4\pi d/L}{\sinh(4\pi d/L)}\right] = nC,$$ (5.51)

where

$$n = \frac{1}{2}\left[1 + \frac{4\pi d/L}{\sinh(4\pi d/L)}\right].$$ (5.52)

In deep waters, the term $\left(4\pi d/L\right)\sinh\left(4\pi d/L\right)$ is approximately zero and

$$C_g = \frac{1}{2}\frac{L_o}{T} = \frac{1}{2}C_o.$$ (5.53)

The group velocity is one-half of the phase velocity. In shallow water, $\sinh(4\pi d/L) \approx 4\pi d/L$ and C_g is the following:

$$C_g = \frac{L}{T} = C \approx \sqrt{gd}.$$ (5.54)

Therefore, the group and phase velocities are equal. In shallow water, because wave celerity is determined by the depth, all component waves in a wave train will travel at the same speed precluding the alternate reinforcing and canceling of components. In deep and transitional water, wave celerity depends on the wavelength. Therefore, longer waves travel will faster and produce the small phase differences resulting in wave groups. These waves are said to be dispersive or propagating in a dispersive medium. This occurs in a medium where their celerity is dependent on wavelength.

The phase velocity of gravity waves in deep water is greater than the group velocity. If an observer follows a group of waves at group velocity they will see waves that originate at the rear of the group that appear to move forward through the group traveling at the phase velocity and disappear at the front of the group.

5.3.2.7 Wave Energy and Power

The total energy of a wave system is the sum of its kinetic and potential energies. The kinetic energy is due to particle velocities associated with wave motion. Potential energy results from part of the fluid mass being above the trough: the wave crest. According to the Airy theory, if the potential energy is determined relative to still water level (SWL), and assuming all waves are propagated in the same direction, potential and kinetic energy components are equal, the total wave energy in one wavelength per unit crest width is given by the following equation:

$$E = E_k + E_p = \frac{\rho g H^2 L}{16} + \frac{\rho g H^2 L}{16} = \frac{\rho g H^2}{8}.$$ (5.55)

Subscripts k and p refer to kinetic and potential energies. Total average wave energy per unit surface area, termed the specific energy or energy density, is then given by

$$\bar{E} = \frac{E}{L} = \frac{\rho g H^2}{8}.$$
(5.56)

Wave energy flux is the rate at which energy is transmitted in the direction of wave propagation across a vertical plane perpendicular to the direction of wave advance and extending down the entire depth. The average energy flux $(\bar{P})$ per unit wave crest width transmitted across a vertical plane perpendicular to the direction of wave advance is therefore the following:

$$\bar{P} = \bar{E}nC = \bar{E}C_g.$$
(5.57)

Energy flux $(\bar{P})$ is frequently called wave power and

$$n = \frac{1}{2}\left[1 + \frac{4\pi d/L}{\sinh\left(4\pi d/L\right)}\right].$$
(5.58)

If a vertical plane is other than perpendicular to the direction of wave advance, then $P = E\, C_g \sin \varphi$, where φ is the angle between the plane across which the energy is being transmitted and the direction of wave advance.

For deep and shallow water Equation (5.55) becomes

$$\bar{P}_o = \frac{1}{2}\bar{E}_o C_o \ \text{(deep water)},$$
(5.59)

$$\bar{P} = \bar{E}C_g = \bar{E}C \ \text{(shallow water)}.$$
(5.60)

An energy balance for a region through which waves are passing will reveal that, for a steady state, the amount of energy entering the region will equal amount leaving the region provided no energy is added or removed from the system. Therefore, when the waves are moving so that their crests are parallel to the bottom contours,

$$\bar{E}_o n_o C_o = \bar{E}nC.$$
(5.61)

or since $n_o = \frac{1}{2}$

$$\frac{1}{2}\bar{E}_o C_o = \bar{E}nC.$$
(5.62)

5.3.2.8 Wave Heights in Deep Water

The size of a wave (i.e., its height and wavelength) in deep water is determined not only by the force of the wind, but also by its duration and by the distance over which

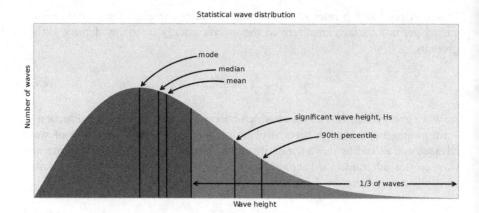

FIGURE 5.10 Statistical wave distribution.

(Wikipedia, Creative Commons CC0 License)

it blows, or its fetch. The significant wave height (SWH or Hs) is defined traditionally as the mean wave height (trough to crest) of the highest third of the waves (H1/3) (Figure 5.10).

5.4 STOKES WAVE THEORY

An extension of the Airy theory to waves of finite height was made by G.G. Stokes in 1846. His method was to expand the wave solution in series form and determine the coefficients of the individual appropriate hydrodynamic equations for finite-amplitude waves.

Stokes wave theory (second, third, or fifth order) is valid for nonlinear waves on intermediate and deep water. That is, for wavelengths (L) that are not large compared to water depth (h). In shallow water, the low-order Stokes expansion breaks down. That is, it tends to give unrealistic results.

Stokes carried his analysis forward to third order of accuracy in the wave steepness H/Δ. This solution has been presented by Skjelbreia (1958) and Wiegel (1954). An extension of the method to fifth order has been made by Skjelbreia and Hendrickson (1962). This work is typically referred to as Stokes fifth-order wave theory and is used for finite-amplitude waves.

Due to numerical difficulties the theory is considered valid for water where the relative depth h/ΔΔ is greater than 1/6.

For a wave of height H, wave number k, and frequency w propagating in the positive x-direction, the free-surface water deflection η from the still-water level is, according to the Stokes fifth-order theory, given by Equation (5.63).

$$\Delta = \frac{1}{k} \sum_{n=1}^{5} Fn \cos n(kx - wt), \qquad (5.63)$$

where F_1, F_2, F_3, etc. are given by the following:

$$F_1 = a$$
$$F_3 = a^2 F_{22} + a^4 F_{24}$$
$$F_3 = a^3 F_{33} + a^5 F_{35} \tag{5.64}$$
$$F_4 = a^4 F_{44}$$
$$F_5 = a^5 F_{55}$$

with F_{22}, F_{24}, etc. denoting wave-profile parameters dependent on kH and parameter a denoting a wave height parameter related to the wave height through Equation (5.65).

$$kH = 2(a + a^3 F_{33} + a^5(F_{35} + F_{55})). \tag{5.65}$$

The horizontal water velocity u and the vertical water velocity v (at place x, time t, and distance y above the seafloor) caused by the free-surface wave propagating over water of depth h are expressible as

$$u = \frac{w}{k} \sum_{n=1}^{5} Gn \frac{\cosh n k y}{\sinh n k h} \cos n (kx - wt), \tag{5.66}$$

$$v = \frac{w}{k} \sum_{n=1}^{5} G_n \frac{\sinh nky}{\sin k \ nkh} \sin n (kx = wt), \tag{5.67}$$

where:

$G_1 = a\, G_{11} + a^3\, G_{13} + a^5\, G_{15}$
$G_2 = 2\,(a^2\, G_{22} + a^4\, G_{24})$
$G_3 = 3\,(a^3\, G_{33} + a^5\, G_{35})$
$G_4 = 4\, a^4\, G_{44}$
$G_5 = 5\, a^5\, G_{55}$
G_{11}, G_{13}, etc. denote wave velocity parameters dependent on kh.

Explicit expressions for F_{22}, F_{24}, G_{11}, etc. are given by Skjelbreia and Hendrickson (1962).

In addition to the above it is also necessary to have the frequency relation connecting wave frequency with wave number.

$$w = gk\,(1 + a^2\, C_1 + a^4\, C_2)\, \tanh\, kh, \tag{5.68}$$

where C_1 and C_2 are frequency parameters.

The wave speed C is determined as in the Airy theory from the relation $C = w/k$, which for the Stokes fifth-order solution is expressible as follows:

$$C = \left(\frac{g}{k}(1 + a^2\, C_1 + a^4\, C_2)\, \tanh kh \right)^{\frac{1}{2}}. \tag{5.69}$$

The gage pressure (p) in the water resulting from the overhead wave and the hydrostatic contribution can also be determined from the velocity components by substitution into the following equation.

$$p = p_w \frac{w}{k} u - \frac{1}{2} r \left(u^2 + v^2 \right) = \frac{\rho_w g}{k} \left(a^2 C_3 + a^4 C_4 + ky' \right), \tag{5.70}$$

where $y' = y - h$; C denotes pressure parameters dependent on kh or h/L.

5.5   CNOIDAL WAVE THEORY

For shallow water, Cnoidal wave theory is generally regarded as being more correct than the Stokes wave theory. This theory was first presented by Korteweg and de Vires in 1895 and has since been developed further by several writers. Wiegel (1960) has summarized these developments and has given a presentation of the theory convenient for practical application.

Cnoidal waves are periodic waves with surface profiles (h) are described in terms of the wave number k and frequency ω as given by Equation. (5.71).

$$h = h_T + H \, cn^2 (kx - \Delta t, m). \tag{5.71}$$

5.5.1   FETCH AND SEAS

As the wind begins to blow over the surface of the ocean an amount of energy is imparted to form wind waves. The first waves to form are capillary waves. These waves in turn, make the sea somewhat rougher and allow more efficient interaction between the wind and the sea surface and therefore allows a more efficient transfer of energy from the wind to the sea.

The harder the wind blows, the greater the amount of energy transfer the larger the waves. The size of the resulting waves (i.e., height and wavelength) is a function of the wind force, its duration, and the distance over which it blows (i.e., fetch). The minimum fetch and duration required for full development of waves associated with various wind speeds is present in Table 5.3.

TABLE 5.3
Full Wave Development Requirements

Wind Speed m/s (kt)	Fetch (km)	Duration (hr)
6.1 (10)	16.5	2.4
6.2 (20)	140	6
16.3 (30)	520	23
20.4 (40)	1320	42
26.5(50)	2570	66

Note: Data from Wolfe et al. (1966).

TABLE 5.4

Characteristics of Fully Developed Wind Waves

Wind Speed m/s (kt)	Average Period (s)	Average Length (m)	Average Height (m)	Maximum Height (m)	Approximate Celerity m/s (kt)
6.1 (10)	2.9	6.5	0.27	0.55	6.6 (9)
6.2 (20)	6.7	32.9	1.5	3.0	6.7 (17)
16.3 (30)	6.6	76.5	6.1	6.5	13.3 (26)
20.4 (40)	6.4	136.0	6.5	16.3	16.8(35)
26.5 (50)	16.3	212.0	16.8	30.0	21.9 (43)

Note: Data from Wolfe et al. (1966).

A fully developed wave is the limiting condition. The characteristics of fully developed waves are presented in Table 5.4. A review of Table 5.4 indicates that the sea surface away from the storm has a smoother appearance.

5.6 SEA STATE

A sea state in oceanography is the condition of the free surface on a large body of water with respect to wind waves and swell at a specific location and time. A sea state is characterized by wave height, period, and power spectrum. The sea state varies with time, as the wind conditions or swell conditions change. The sea state can be determined by an experienced observer, or through instruments such as weather buoys, wave radar, or remote sensing satellites. Because of the large number of variables involved in determining the sea state it cannot be quickly or easily determined. Therefore simpler scales are normally used to give an approximate but accurate description of sea conditions. The most frequently used scale is the World Meteorological Organization (WMO) sea state code, refer to Table 5.5. In addition,

TABLE 5.5

World Meteorological Organization Sea State Code

Sea State	Wave Height (m)	Characteristics
0	0	Calm (glassy)
1	0–0.1	Calm (rippled)
2	0.1–0.5	Smooth (wavelet)
3	0.5–1.25	Slight
4	1.25–2.5	Moderate
5	2.5–4	Rough
6	4–6	Very Rough
7	6–9	High
8	9–14	Very High
9	Over 14	Phenomenal

TABLE 5.6
Character of the Sea Swell

	0 None
Low	1. Short or average
	2. Long
Moderate	3. Short
	6. Average
	6. Long
High	6. Short
	6. Average
	6. Long
	6. Confused

Note: Direction from which swell is coming
         should be indicated.

the character of sea swells can be described using Table 5.6. The WMO sea state code largely adopts the "wind sea" definition of the Douglas Sea State.

5.7   STORM SURGES

5.7.1   INTRODUCTION

A storm over near shore waters can generate large water level fluctuations if it is sufficiently strong and the region is shallow over a large enough area. This is commonly known as a storm surge or meteorological tide. Storm activity can cause both a set up (rise) and set down (fall) of the water level at different locations and times, with the set up predominating in magnitude, duration, and areal extent. Specific causes of water level change include the following: (1) surface wind stress, (2) Coriolis acceleration, (3) long wave generation by a moving pressure disturbance, (4) atmospheric pressure differentials, and (5) precipitation and surface runoff.

Storm surge calculations require knowledge of the spatial and temporal distribution of wind speed, wind direction, and surface air pressure for the design storm conditions.

5.7.2   WIND FASTEST MILE

The greatest wind speed to be expected at a particular site can be estimated from analysis of local daily weather reports. Due to this fluctuation in wind speed records, they are averaged over the time required for a horizontal column of air 1 mile long to pass the measuring station. The fastest mile of wind is then defined as the highest wind speed so measured in a single day. The annual extreme fastest mile of wind is the largest of the daily maximums recorded during a single year.

Statistical projections of the gust wind speeds 30 feet above the earth that can be expected are shown in Table 5.7.

TABLE 5.7
Gust Speeds for Mean Hourly Speeds Between 20 and 80 Knots

Duration of Gust	G_{30}/V_1	Reference
1.0 minutes	1.25	Bretschnieder (1969)
5 seconds	1.48	Bretschnieder (1969)
0.5 seconds	1.61	Bretschnieder (1969)
5 seconds	1.2	Gentry (1953)
0.5 seconds	1.3	Gentry (1953)

Notes: (1) G_{30} is the gust speed at 9.1 m (30 feet) above the surface; (2) V_1 is the 1 minute average wind speed.

5.7.3 GUST VELOCITY

Gustiness depends upon the temperature gradient, the mean wind speed, and altitude. Gusts can occur in both horizontal and vertical directions. At heights below 30.5 m (100 feet), the velocity of horizontal gusts is larger than vertical gusts. At heights of 30.5 m (100 feet) or more above the surface, vertical gusts can be of the same order of magnitude as the horizontal. Gust factors are defined as the ratio of gust speed at 9.1 m (30 feet) above the surface to the 1 minute average wind speed have been proposed by Bretschnieder (1969) and Gentry (1953). These values are summarized in Table 5.7. Gust factors increase with elevation above the surface of the sea as given in Equation (5.72).

$$\frac{G_z}{G_{30}} = \left(\frac{Z}{30} \right)^{\frac{1}{12}},$$
(5.72)

where:

G_z—gust speed at elevation Z
Z—elevation above surface of water
G_{30}—gust speed at 9.1 m (30 feet) above the surface of the water

Wind speeds in general can be corrected to any duration using the method presented by Vellozzi and Cohen (1968).

5.8 TSUNAMIS

5.8.1 INTRODUCTION

Tsunamis or seismic sea waves are impulsively generated, dispersive waves of relatively long period and low amplitude. These waves are typically generated by sudden large-scale sea floor movements usually associated with severe, shallow focus earthquakes. Typically an earthquake of at least 6.5–7.5 Richter magnitude and with focal depths of less than 48–64 km is required to initiate such sea floor movements.

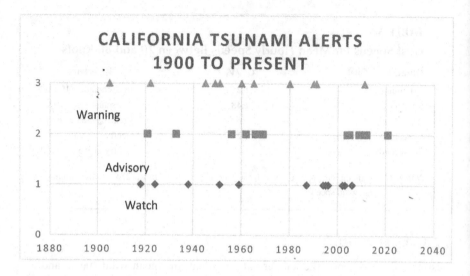

FIGURE 5.11   Likely tsunami alerts since 1900 based on present-day definitions.
(Adapted from Dengler (2022))

Tsunamis may also be generated by underwater landslides, volcanoes, or explosions. The number of likely tsunami alerts since 1910 based on present-day definitions is presented in Figure 5.11.

5.8.2 TSUNAMI GENERATED BY EARTHQUAKES

Since tsunamis are associated with seismic activity the most destructive cases have been recorded in the North Pacific Ocean rim of fire. Susceptible areas of engineering importance are typically associated with low-lying coastal regions. Historical records of tsunamis damage have been discussed by Bascom (1980) and there have been a number engineering investigations of tsunamis (Magoon, 1965; Matlock et al., 1962; Murty, 1977; Wilson and Torum, 1968). Tsunamis are categorized as long waves (wavelength of 16.1 km (100 miles or more)), therefore, travel times can be estimated by using either maps or for deep water a simple mathematical relationship can be utilized. Maps do not provide information on the height or the strength of the wave, only the arrival times. Due to the large wavelength tsunamis can travel everywhere in the sea at approximately the shallow water wave speed as given by Equation (5.73).

$$C = \sqrt{gd},\tag{5.73}$$

where:

C = wave celerity or velocity
g = acceleration due to gravity
d = water depth

Assuming an average depth for the Pacific Ocean of 12,000 feet, then a tsunami wave will travel at approximately 400 knots (knots × 1.8532 = km/hr). Then if the location of the epicenter of the earthquake is known, the travel time of a tsunami between points can be estimated either using nautical charts by summing the incremental travel times along a wave front. The time it takes a tsunami to travel a given distance is estimated using Huygen's principle. This principle states that all points on a wave front are point sources for secondary spherical waves. Minimum travel times are computed over a grid starting at the earthquake epicenter. Times are then computed to all surrounding grid points from the starting point. The grid point with minimum time is then taken as the next starting point and times are computed from there to all surrounding points. The starting point is continually moved to the point with minimum total travel time until all grid points have been evaluated. This technique is discussed in the following (Shokin et al., 1986).

There are a number of situations in which the estimated arrival times may not agree with the observed arrival times of the tsunami waves. This is due in part to the following:

- Bathymetry is not accurate in the vicinity of the epicenter.
- Epicenter is not well located, or its origin time is uncertain.
- Epicenter is on land and a pseudo epicenter off the coast must be selected.
- Bathymetry is not accurate in the vicinity of the reporting station.
- Nonlinear propagation effects may be important in shallow water.
- Observed travel times do not represent the first wave but instead are later arrivals.

The damage due to a tsunami is normally from large hydrostatic and hydrodynamic forces along with the impact of water-borne objects, overtopping with subsequent flooding and erosion caused by the high water velocities. The characteristics of tsunamis of interest are primarily those associated with the nearshore environment. These characteristics are (1) run-up heights, (2) surge or bore velocities, and (3) return period.

5.8.2.1 Crescent City, California

Since 1933, 31 tsunamis have been observed in Crescent City. Four of those caused damage, and one of them, in March 1964, remains the "largest and most destructive recorded tsunami to ever strike the United States Pacific Coast" (University of Southern California's Tsunami Research Center). The 1964 tsunami killed 17 people on the West Coast, 11 of them in Crescent City.

5.8.2.2 1964 Alaskan Earthquake

The 1964 tsunami was caused by the largest earthquake ever recorded in North America. The so-called Good Friday Earthquake struck during late afternoon, its epicenter was just north of Alaska's Prince William Sound, registered 9.2 on the Richter scale and killed 115 Alaskans, inflicting its worst damage on Anchorage. All but nine of the deaths were caused not by the earthquake itself, but by the tsunami that resulted. (The tsunami also hit Canada, but no one there was killed.)

TABLE 5.8

Tsunamis That Have Hit Crescent City in the Recent Past

Year	Wave Height (ft.)	Earthquake Causing Earthquake
1960	3 (?)	Chilean EQ, Mag. 9.5
1964	21	Alaskan EQ, Mag. 9.2
2006	5.8	Kuril Islands, Japan, Mag. 8.3
2011	6.5	Fukushima, Japan, Mag. 9

5.8.2.3   Tsunami Damage and Loss of Life

After Alaska, California, where the tsunami hit a little before midnight, was the state worst-hit by the 1964 wave. Total property damage there was $17 million (Oregon and Washington each sustained less than $1 million), of which fully $15 million occurred in Crescent City. Although the earthquake killed many more people in Alaska than in Crescent City, the property damage per block ended up, being greater in Crescent City. A listing of recent tsunamis to hit Crescent City is presented in Table 5.8.

Crescent City is presumed to be more vulnerable to tsunamis than any other city along the West Coast of the United States, based on frequency of impact from past events. Tsunami waves tend to get amplified in the area around Crescent City, and the observed wave heights in Crescent City Harbor are typically an order of magnitude greater than those measured in other locations along the West Coast. The reasons for amplification of tsunami waves have been clearly identified, though most evidence points to the combined effect of two factors:

1. The presence of the Mendocino Escarpment, an abrupt 1000-m seafloor depth discontinuity immediately offshore of the Northern California coast, with the potential for channeling tsunami energy toward Crescent City.
2. The tendency of the Crescent City Harbor to amplify wave frequencies around the 20-minute period, perhaps being the most likely cause of elevated tsunami wave heights observed in the area. In addition, the combination of geology and tectonics has thrust Point St. George out into the ocean and created a natural south facing crescent-shaped bay that funnels tsunamis from all directions into the harbor area (Dengler et al., 2015).

A number of studies have been conducted to look at the potential flooding due to tsunami run-up in Del Norte County, CA, Oregon and Washington States (Arcas and Uslu, 2010; California Geological Survey [CGS] 2009, 2021a, 2021b; Redwood Coast Tsunami Working Group [RCTWG] 2023; Uslu et al., 2007). These different groups have produced a series of potential tsunami inundation maps for various cities (i.e., Humboldt Bay, California; Crescent City, California; Brookings, Oregon; Coos Bay/North Bend, Oregon; Newport, Oregon; Nehalem, Oregon; and Grays Harbor, Washington State) along their respective shorelines. These tsunami inundation maps are presented in Figures 5.12–5.18.

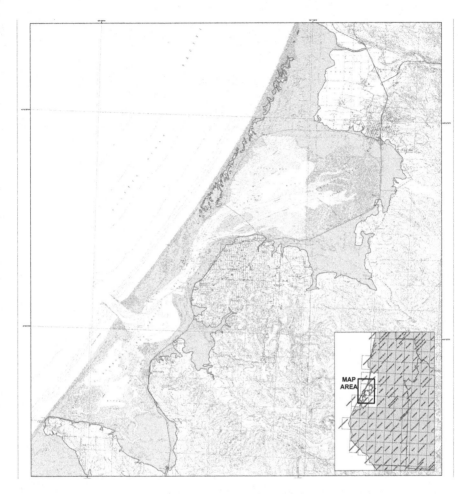

FIGURE 5.12 Tsunami inundation map for emergency planning, Humboldt Bay, California. Inundation indicated by darker color adjacent to water.

(Courtesy of the California Geological Survey, June 1, 2009)

5.8.3 MECHANISM OF TSUNAMI GENERATION

Tsunamis or seismic sea waves are impulsively generated, dispersive waves of relatively long period and low amplitude. These waves are typically generated by sudden large-scale sea floor movements usually associated with severe, shallow focus earthquakes. Usually an earthquake of at least 6.5–7.5 Richter magnitude and with focal depths of less than 48–64 km is required to initiate such sea floor movements (Figure 5.19). Tsunamis may also be generated by underwater landslides, volcanoes, or explosions. An example of a large-scale sea floor movement is a simple fault in which tension in the basement rock is relieved by the abrupt rupturing of the rock along an inclined plane as shown in Figure 5.19e. When such a fault occurs, a large mass of rock and sediment drops rapidly and the support is consequently removed from a column of water that extends to the surface. The water surface oscillates up

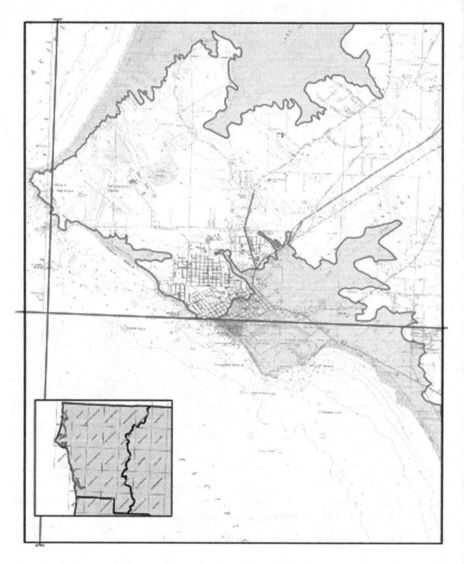

FIGURE 5.13 Tsunami inundation map for emergency planning, Crescent City, California. Inundation indicated by darker color adjacent to water.

(Courtesy of the California Geological Survey, March 20, 2009)

and down as it seeks to return to mean sea level, and a series of waves are produced. In contrast, a tsunami can also be generated if the basement rock fails in compression, the mass of rock on one side rides up and over that of the other, as shown in Figure 5.19c and a column of water is lifted. Another mechanism is a landslide or mass movement which is initiated by an earthquake (Figure 5.19h). If the slide begins above the water, abruptly dumping a mass of rock and soil into the sea, waves are generated. If the slide occurs well below the surface of the sea, it also can create waves. Tsunamis can be highly destructive waves, especially at certain locations

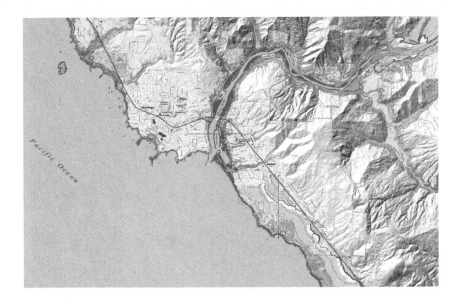

FIGURE 5.14 Tsunami inundation map for emergency planning, Brookings, Oregon. Inundation indicated by darker color adjacent to water.

(Courtesy of the State of Oregon, Department of Geology and Mineral Industries, 2012)

FIGURE 5.15 Tsunami inundation map for emergency planning, Coos Bay/North Bend, Oregon. Inundation indicated by darker color adjacent to water.

(Courtesy of the State of Oregon, Department of Geology and Mineral Industries, 2012)

FIGURE 5.16 Tsunami inundation map for emergency planning, Newport, Oregon. Inundation indicated by darker color adjacent to water.

(Courtesy of the State of Oregon, Department of Geology and Mineral Industries, 2012)

FIGURE 5.17 Tsunami inundation map for emergency planning, Nehalem Bay, Oregon. Inundation indicated by darker color adjacent to water.

(Courtesy of the State of Oregon, Department of Geology and Mineral Industries, 2012)

FIGURE 5.18 Tsunami inundation map for emergency planning, Grays Harbor, State of Washington. Inundation on all flat areas adjacent to water.

(Courtesy of the State of Oregon, Department of Geology and Mineral Industries, 2012)

prone to tsunami run-up. Although they are almost undetectable at sea, because of their long wavelengths, with periods of a few minutes to an hour or more, and heights of only 0.3 m or 0.6 m or less, when they approach shallow water, shoaling, refraction, and possible resonant effects can cause run-ups from several meters to upward of approximately 30 m or more, depending on the tsunamis characteristics and the local typography. Tsunamis are often observed as a series of highly periodic surges that may continue over a period of several hours.

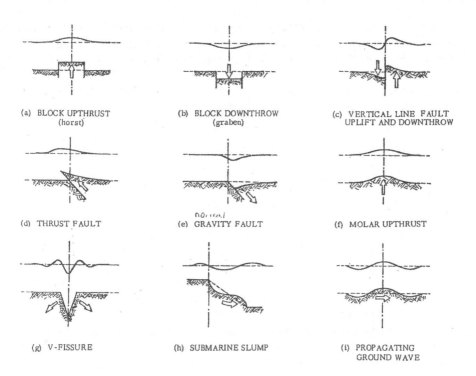

FIGURE 5.19 Schematic representations of tectonic movements and initial water surface disturbances likely to occur in submarine earthquake. (a) Block upthrust (horst), (b) block downthrow (graben), (c) vertical line fault (uplift and downthrow), (d) thrust fault, (e) gravity fault, (f) molar upthrust, (g) v-fissure, (h) submarine slump, and (i) propagative ground wave.

(Wilson, 1969. Courtesy of the Naval Civil Engineering Laboratory, Port Huene, California)

Since tsunamis are associated with seismic activity the most destructive cases have been recorded in the North Pacific Ocean rim of fire. Susceptible areas of engineering importance are typically associated with low-lying coastal regions. Historical records of tsunamis damage have been discussed by Bascom (1980) and there have been a number engineering investigations of tsunamis (Magoon, 1965; Matlock et al., 1962; Murty, 1977; Wilson and Torum, 1968). The height characteristics of some recent tsunamis are shown in Table 5.9. Tsunamis are categorized as

TABLE 5.9
Highest Recently Recorded Tsunamis

Year	Height (m)	Location
1958	524	Lituya Bay
1963	250	Vajont Dam
1980	260	Spirit Lake
1964	70	Alaska Earthquake

long waves (wavelength of 160 km or more), therefore, travel times can be estimated by using either maps or for deep water a simple mathematical relationship can be utilized. Maps do not provide information on the height or the strength of the wave, only the arrival times.

There are a number of situations in which the estimated arrival times may not agree with the observed arrival times of the tsunami waves. This is due in part to the following:

- Bathymetry is not accurate in the vicinity of the epicenter.
- Epicenter is not well located, or its origin time is uncertain.
- Epicenter is on land and a pseudo epicenter off the coast must be selected.
- Bathymetry is not accurate in the vicinity of the reporting station.
- Nonlinear propagation effects may be important in shallow water.
- Observed travel times do not represent the first wave but instead are later arrivals.

The damage due to a tsunami is normally from large hydrostatic and hydrody-namic forces along with the impact of water-borne objects, overtopping with subse-quent flooding and erosion caused by the high water velocities. The characteristics of tsunami of interest to engineers are primarily those associated with the nearshore environment. These characteristics are (1) run-up heights, (2) surge or bore velocities, and (3) return period.

5.9 SUMMARY

The most important factor in coastal development are probably waves. The wave activity can cause erosion but also transport sediment either directly or through the generation of wave-driven currents. The difficulty in handling waves is describing the relationship between fluid motion (i.e., waves) and a solid sub-strate. This topic is approached by looking at the basic hydrodynamic equations governing wave motion. These are equations describing momentum and continuity for an incompressible irrotational inviscid fluid with constant density along with the appropriate boundary conditions. In this chapter, gravity waves are considered. The emphasis is on linear wave theory but Stokes and Cnoidal waves are intro-duced. In addition, the issue of tsunamis is presented. This topic is approached by looking at the mechanism of generation by earthquakes and the resulting damage and loss of life.

REFERENCES

Arcas, D., and Uslu, B. (2010). PMEL Tsunami Forecast Series: Vol. 2 A Tsunami Forecast Model for Crescent City, California, NOAA OAR Special Report, March.
Bascom, W. (1980). *Waves and Beaches*. New York: Anchor Books.
California Geological Survey (CGS). (2009). Crescent City Inundation Map for Emergency Planning.
California Geological Survey (CGS). (2021a). Information Warehouse: Tsunami Hazard Area Maps. https://maps.conservation.ca.gov/cgs/informationwarehouse/ts_evacuation/

California Geological Survey (CGS). (2021b). Tsunami Web App. https://rctwg.humboldt. edu/tsunami-hazard-maps

Chaney, R. C. (2021). *Marine Geology and Geotechnology of the South China Sea and Taiwan Strait.* CRC Press.

Dengler, L. (2022). California's tsunami preparedness week approaches. *Time Standard Newspaper,* March 13.

Dengler, L., Barberopoulou, A., Uslu, B., and Yim, S. (2015). Tsunami damage in Crescent City, California from the November 15, 2006 Kuril event. Research Gate.

Dengler, L., Goltz, J., Fenton, J., Miller, K., and Wilson, R. (2021). Building tsunami-resilient communities in the United States: An example from California. *TsuInfo Alert,* 13(2). Washington State Department of Natural Resources.

Eagleson, P. S., and Dean, R. G. (1966). Small amplitude wave theory. In Ippen, A.T. (Ed.), *Estuary and Coastline Hydrodynamics.* New York, NY: McGraw-Hill Book Co., 1–95.

Eckart, C. (1952). Propagation of gravity waves from deep to shallow water. Gravity Waves. National Bureau of Standards Circular 521, U.S. Govt. Print. Off., Washington, DC.

Gentry, R. C. (1953). Wind velocities during hurricanes. *Proceedings American Society of Civil Engineers.* 25 pp.

Kinsman, B. (1965). *Wind Waves.* Englewood Cliffs, NJ: Prentice Hall, Inc.

Magoon, O. T. (1965). Structural Damage by tsunamis, Coastal Engineering Specialty Conference, Santa Barbara, CA, October, American Society of Civil Engineers, pp. 35–68.

Matlock, H., Reese, L., and Matlock, R. R. (1962). Analysis of Structural damage from the 1960 tsunami at Hilo, Hawaii, Technical Report Defense Atomic Support Agency 1268, Structural Mechanics Research Lab, University of Texas, Austin, March: 95 pp.

McCormick, M. E. (1973). *Ocean Engineering Wave Mechanics.* New York, NY: Wiley.

Murty, T. S. (1977). Seismic sea waves Tsunamis, Bulletin of the Fisheries Research Board of Canada, Toronto, Canada, 337 pp.

Redwood Coast Tsunami Working Group (RCTWG). (2023). https://rctwg.humboldt.edu

Shokin, et al. (1986). Calculations of tsunami travel time charts in the Pacific Ocean. *Science of Tsunami Hazards,* 5, 85–113.

Skjelbreia, L. (1958). *Gravity waves, Stokes third wave order.* Council on Wave Research, The Engineering Foundation of the California Research Corporation.

Skjelbreia, L., and Hendrickson, J. A. (1962). Fifth order gravity wave theory and tables of functions. National Engineering Science Company.

Uslu, B., Borrero, J. C., Dengler, L., Tsunami, C. E., and Synolakis, C. E. (2007). Inundation at Crescent City, California generated by earthquakes along the Cascadia subduction Zone. *Geophysical Research Letters,* 34(20). 1–5.

Vellozzi, J., and Cohen, E. (1967). Gust response factors for buildings and other structures. Conference Preprint # 434, Amer. Soc. Civil Engs., Environmental Engineering Conf., Dallas, TX, 31 pp.

Wiegel, R. L. (1954). Oscillatory Waves. U.S. Army, Beach Erosion Board, Bulletin, Special Issue No. 1, July.

Wiegel, R. L. (1960). Transmission of waves past a rigid vertical thin barrier. *Journal of the Waterways and Harbors Div., ASCE,* 86(WW1), 1–12.

Wilson, B. W. (1969). Earthquake Occurrence and Effects in Ocean Areas (U). U.S. Naval Civil Engineering Laboratory, Port Hueneme, California, Technical Report Cr. 66.026.

Wilson, B. W., and Torum, A. (1968). The Alaskan Tsunami of March 1964: Engineering Evaluation Technical Memorandum No. 25, Coastal Engineering Research Center, Corps of Engineers, U.S. Army, Washington, DC..

Wolfe, J. H., Silva, R. W., Mckibbin, D. D., and Matson, R. H. (1966, July 1). The compositional, anisotropic, and nonradial flow characteristics of the solar wind. *The Journal of Geophysical Research,* 1(23), 3329–3335.

6 Wave Behavior in Shallow Water and Run-Up

6.1 INTRODUCTION

When a wave encounters a coastal structure or beach, a process begins to dissipate energy. The effects of waves on the coastal zone can be divided into three zones which are (1) offshore zone, (2) shoaling zone, and (3) surf zone. The surf zone has been described by Barua (2016) as the shoreline region exists from the initial wave breaking seaward to the still water level shoreward. This zone shifts continuously from shoreward during high tide to seaward during low tide. In addition, the zone shifts shoreward during low waves and then to seaward during high waves. To help visualize the coastal zone an idealized beach profile is presented in Figure 6.1. A review of this figure defines the various elements of the coastal area. The beach configuration changes with the amount of wave energy being dissipated. This configuration of the beach changes is reflected between winter storm waves and the summer more moderate wave climate.

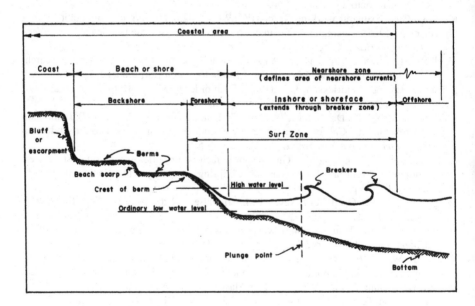

FIGURE 6.1   Beach profile nomenclature.

(U.S. Army Coastal Engineering Research Center, 1977. Courtesy of the Department of the Army COE)

   DOI: 10.1201/9781003454212-8

During the winter, storm waves remove beach material from the shore and deposit it in a bar offshore. In contrast, during the more moderate wave climate of summer material is replaced on the shore thus building up the beach. This process is shown schematically in Figure 6.2.

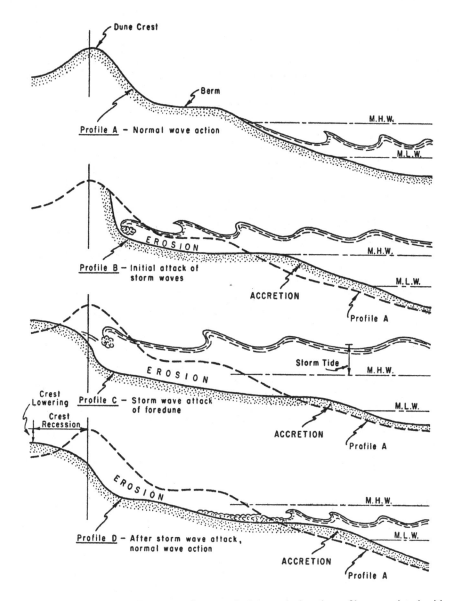

FIGURE 6.2   The general pattern of seasonal changes in beach profiles associated with parallel variations in wave energies.

(U.S. Army Coastal Engineering Research Center, 1977. Courtesy of the Department of the Army COE)

The effectiveness of waves to mobilize sediment depends in large part on whether it can feel the bottom. The relationship between wave energy to sediment mobility and coastal morphology is illustrated in Figure 6.3. A review of Figure 6.3 shows how the extent of wave energy dissipation will control whether the coast becomes

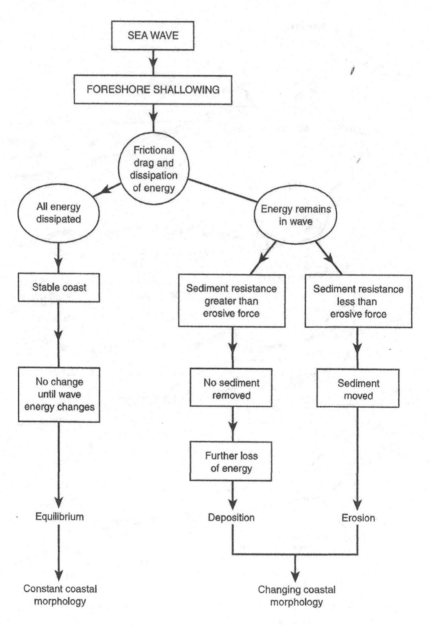

FIGURE 6.3 Relationship of wave energy to sediment mobility and coastal morphology. (French, 1997; republished with permission of Taylor & Francis)

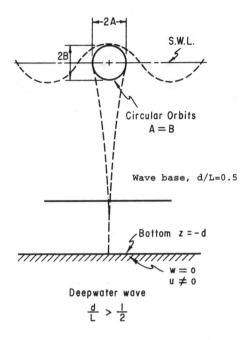

FIGURE 6.4   Idealized orbits of water particles in deep water showing the wave base.

(U.S. Army Coastal Engineering Research Center, 1977. Courtesy of the USA-COE)

erosion, depositional, or stable. The movement of sediment depends on the depth that the movement of water particles extends. The effect of water waves to move bottom sediments in deep water extends to a depth (d) related to the wavelength ($d = L/2$) as shown in Figure 6.4. This depth is called the wave base. If the water is deeper than the wave base ($d \geq L/2$) the orbits of water particles become circular. The wave base can be defined as the depth to which a surface wave can move water particles. If the depth of water extends beyond the wave base then there is no interaction between the seabed and the wave. As the wave starts undergoing shoaling the water particles start to feel the bottom. When water waves approach a coastline their water particle orbits undergo a change from circular to elliptical. This change is due to whether the wave base is the depth to which a surface wave can move water as shown previously in Figure 6.4.

If the water is shallower than the wave base the orbits of the water particles become more elliptical as they interact with the seabed, Figure 6.5. The elliptical orbits of the water particles become progressively flatter as the seabed rises. As the wave approaches the coast, its height increases and wavelength decreases as the depth decreases.

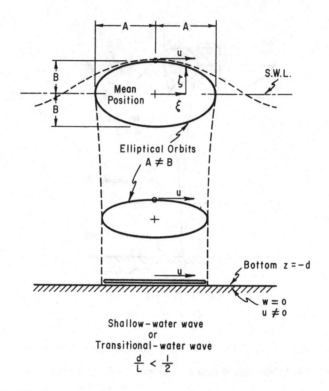

FIGURE 6.5   Idealized orbits of water particles in shallow water.

(U.S. Army Coastal Engineering Research Center, 1977. Courtesy of the USA-COE)

6.2   WAVE BREAKING

Wave shoaling, refraction, and diffraction transform the waves from deep water to the point where they break and then the wave height begins to decrease markedly, because of energy dissipation. The sudden decrease in the wave height is used to define the breaking point and determines the breaking parameters (H_b, d_b, and x_b). The wave steepness (H/L) associated with its breaking is defined as the ratio of wave height (H) divided by wavelength (L). When the wave steepness becomes ($H/L \geq 1/7$), it becomes unstable and breaks (i.e. surf zone). Four types of breaking waves occur in the surf zone on the coast depending on the following: (1) incident wave characteristics and (2) bottom slope (Galvin Jr., 1968). These breaking waves are spilling, plunging, collapsing, and surging (refer to Figure 6.6).

The type of breaking wave depends on the Iribarren number (ξ_i) or surf similarity parameter. The surf similarity parameter compares the wave surface slope to the bed slope (m), the still water depth h, the significant incident water height H_s, and the wave period T, where all quantities refer to a location just seaward of the breakpoint.

$$\xi_i = \frac{m}{\sqrt{H_s/L}}, \tag{6.1}$$

a) Spilling breaking wave

b) Plunging breaking wave

c) Surging breaking wave

d) Collapsing breaking wave

FIGURE 6.6 Types of breaking waves.

(From U.S. Army Coastal Engineering Research Center, 1984. Courtesy of the USA-COE)

where

$$L = \frac{gT^2}{2\pi}$$ —wavelength at the seaward boundary of the breaker zone

$${H_s}/{L}$$ —wave steepness

Using the expression for the shallow water celerity $(c \approx \sqrt{gh})$ and the condition of surf zone saturation (i.e. all waves are broken and decay in proportion to depth $\gamma = {H_s}/{h}$)

$$\xi_i = \sqrt{\frac{\gamma}{2\pi}} \left(\frac{m}{{H_s}/{L}} \right). \tag{6.2}$$

For periodic waves propagating on a plane beach there are two possible choices depending on the application for the Iribarren number (ξ) as given in Equations (6.3) and (6.4). The reflection coefficient for a reflective object depends on the slope angle α, surface roughness, porosity, and the incident wave steepness H_i/L (Battjes, 1974).

$$\xi_o = \frac{\tan(\alpha)}{\sqrt{{H_o}/{L_o}}}, \tag{6.3}$$

$$\xi_b = \frac{\tan(\alpha)}{\sqrt{{H_b}/{L_b}}}, \tag{6.4}$$

TABLE 6.1
Iribarren Number as Function of
Breaker Type

Breaker Type	Breaking
Spilling	< 0.5
Plunging	$0.5 < \xi_i < 2.5$
Collapsing	$2.5 < \xi_i < 3.7$
Surging	$3.7 < \xi_i$

Source: Adapted from Battjes (1974); Okazaki and Sunamura (1991).

where

H_o—wave height in deep water
H_b—breaking wave height at edge of surf zone
L_o—deep water wavelength

$$L_o = \frac{gT^2}{2\pi}$$

T—period
G—gravitational acceleration
α—angle of the seaward slope of a structure
H/L—wave steepness

The breaker type is dependent on the Iribarren number (ξ_i) (refer to Table 6.1).

There are a number of issues involved when waves travel from deep into shallow water (i.e. shoaling). These issues are illustrated in Figure 6.7 where shoaling, refraction, and diffraction are illustrated. Wave transformation describes the changes that occur in waves as they move from deep into shallow water.

The surf zone is the initial wave breaking to the shoreward limit of the still water level. The extent of this zone shifts continuously—to shoreward during high tide and to seaward during low tide—to shoreward during low waves and to seaward during high waves.

In the following both of these issues will be discussed.

6.3   WAVE REFRACTION

Waves rarely approach the shore at a perfect 90-degree angle. As waves approach the shoreline, they bend so that the wave crests become nearly parallel to the shore. Different segments of the wave crest will travel at different speeds since wave speed in shallow water is proportional to the depth of water. This change in the speed of the wave will cause refraction.

FIGURE 6.7   Wave transformations occurring in Crescent City Harbor, California.

(From Courtesy of Google Earth)

Consider the case of waves moving into shallow water as shown in Figure 6.7. The propagation of ocean surface waves is described by the dispersion relationship. This relationship relates angular frequency (ω) to the wave number (k) and water depth (h). Neglecting surface current effects,

$$\omega = \sqrt{gk \ \tanh(kh)}, \tag{6.5}$$

where

  g—acceleration of gravity
  h—depth of water
  L—wavelength
  T—period
  k—wave number, $k = 2\pi/L$
  ω—angular frequency, $\omega = 2\pi/T$
  $\tanh(kh)$—hyperbolic tangent

In deep water, kh goes to infinity, the hyperbolic tangent $\tanh(kh)$ is then approximately one. Equation (6.5) reduces to the following:

$$\omega = \sqrt{gk}. \tag{6.6}$$

When a wave propagates over the topography of the seabed at some point it will begin to feel the bottom as the depth decreases. When this occurs the wave period (T) remains constant. The dispersion relationship as presented in Equation (6.5) in turn requires that the wavelength (L) becomes shorter, and the phase speed (c) decreases as shown in Equation (6.7).

$$c = \sqrt{\frac{g}{k} \tanh(kh)}. \tag{6.7}$$

This near-shore process of surface gravity waves decreasing celerity is referred to as wave shoaling. Assume a wave partially in both deep and shallow water that is approaching a straight coast at an angle of α_d. Part of this wave that is in deep water will move faster than the part of the wave in shallow water as shown in Equation (6.7). This effect causes the wave to rotate until it is parallel to the bottom contours of the seabed. This process is called wave refraction as shown in Figure 6.8.

Wave height (H) during the refraction process is complex. Wave height H during the shoaling process can be determined using Equation (6.8) (Kinsman, 1965).

$$H = dK_s K_r, \tag{6.8}$$

where

d—the still water depth (SWH) in deep water
K_s—Shoaling coefficient given in Equation (6.9)
K_r—refraction coefficient

$$K_s = \left(\frac{k}{k_d}\right)^{1/2} \left(1 + \frac{2kh}{\sinh(2kh)}\right)^{-1/2}, \tag{6.9}$$

$$K_r = \left(\frac{\cos(\alpha_d)}{\cos(\alpha)}\right)^{1/2}, \tag{6.10}$$

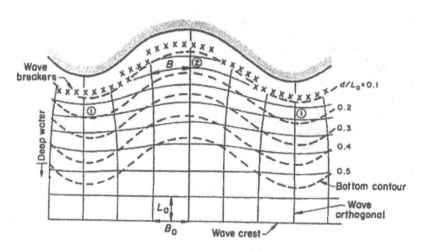

FIGURE 6.8   Refraction diagram.

(From U.S. Army Coastal Engineering Research Center, 1984. Courtesy of the USA-COE)

where k_d—wave number in deep water.

$$K_s = \sqrt{\frac{C_o}{C}} \sqrt{\frac{1}{1 + \dfrac{2kd}{\sinh 2kd}}}, \tag{6.11}$$

$$K_s = \frac{H}{H_o} = \sqrt{\frac{C_{go}}{C_g}}, \tag{6.12}$$

where

C_o—The wave speed in deep water
C—The wave speed at depth d

These two wave speeds can be related by $C_o = \dfrac{\omega}{k_o}$ and $C = \dfrac{\omega}{k}$ where ω is the frequency of the waves, which is the same whether they are in deep or shallow water. Therefore

$$\frac{C_o}{C} = \frac{k}{k_o}. \tag{6.13}$$

And you can use the linear dispersion relation for the frequency at both points:

$$\omega = \sqrt{gk \ \tanh(kd)} = \sqrt{gk_o}. \tag{6.14}$$

$$\frac{k}{k_o} = \frac{1}{\tanh(kd)}. \tag{6.15}$$

By using Snell's law, the parameter α can be shown to be

$$\alpha = \sin^{-1}\left(\frac{k_d}{k} \ \sin(\alpha_d)\right). \tag{6.16}$$

The shoaling effect K_s is a function of only water depth (d). It acts first to decrease the wave height as the waves shoal. The water height increases again when the depth becomes shallow as evidenced by reaching some threshold (i.e. ratio of shallow water depth to wavelength in deep water) lower than 0.056 (Kinsman, 1965). In contrast, the coefficient K_r is taken relative to the approaching angle α_d of the waves in deep water as shown in Equation (6.16). Therefore, a change in wave height during the refraction process is affected by both factors.

The average energy flux in deep water is given as follows:

$$P_o = n_o c_o E_o b_o = \frac{1}{2} c_o \left(\frac{1}{8} \rho g H^2\right), \tag{6.17}$$

where $n_o = \frac{1}{2}$ because short wave depth is larger than $1/2 C_o = C/2$.

The corresponding average energy flux in shallow water is the following:

$$P = ncEb = nc\left(\frac{1}{8}\rho gH^2\right)b. \tag{6.18}$$

Assuming that no power is lost between the wave orthogonals, $P = P_o$.

$$\frac{1}{2}c_o\left(\frac{1}{8}\rho gH^2\right)b_o = nc\left(\frac{1}{8}\rho gH^2\right)b. \tag{6.19}$$

Rearranging

$$\frac{H}{H_o} = \sqrt{\frac{0.5c_o}{cn}}\sqrt{\frac{b_o}{b}}, \tag{6.20}$$

$$\frac{H}{H_o} = K_S K_r, \tag{6.21}$$

where

K_s—shoaling coefficient
K_r—refraction coefficient

A hypothetical shoreline and nearshore bathymetry is shown in Figure 6.8. Consider a series of waves (i.e. wave train) with a deep water wavelength (L_o) that is approaching the shore with a crest orientation parallel to the shoreline. Bottom contour depths or bathymetry are given in terms of the wavelength (L_o) in deep water. As portions of the wave crest enter the area where $d/L_o < 0.5$, the wavelength and celerity begin to decrease. The result is the refraction of the wave train with the wave crest orientation approaching that of the bottom contours. Orthogonal lines that are constructed equally spaced and normal to the wave crests in deep water. Extending these orthogonal lines toward the shore reflects the pattern of energy distribution at any point along the wave crest. Therefore, where orthogonals converge reflects an increase in the energy per unit crest length.

The wave crest height variation that occurs from deep water to the shore is defined by the following:

$$\frac{H}{H_o} = \sqrt{\frac{L_o}{2\pi L}}\sqrt{\frac{B_o}{B}} = \frac{H}{H_o'}\sqrt{\frac{B_o}{B}}. \tag{6.22}$$

The shoaling coefficient ($K_s = H/H_o'$) is only a function of d/L_o in this equation. The orthogonal spacing ratio ($B_o/B = K_r^2$) for a nearshore point is determined from a refraction analysis (Figure 6.8). The calculated change in wave height using Equation (6.22) is the average over the orthogonal spacing B. Refraction in

Figure 6.8 causes a convergence of orthogonals over a submerged ridge (point 1), resulting in higher waves. The presence of this ridge is indicated by the closeness of the contour lines. In contrast, over a submerged trough (point 2) wave heights are lower. This trough is indicated by contour lines being farther apart. They can actually be lower than the deep water height. This can occur if the effect of refraction in lowering the wave height is greater than the increase in wave height owing to shoaling.

Some empirical data for wave processes are presented in terms of the unrefracted deep water wave height H'_o. A wave having a deep water height (H_o) that refracts into a nearshore location where the refraction coefficient is K_r. Then the unrefracted deep water height to be used in these diagrams is $H'_o = K_r H_o$.

The usual objective in constructing a refraction diagram is to evaluate wave height and direction at a particular point or points near shore. To do this, at least two orthogonal lines are needed and they should arrive at the shore bracketing the point of interest. Refraction diagrams were originally constructed manually. They are now typically done by computers. In the following, three different procedures used for both manual and computer analysis will be outlined. These methods are the following: (1) Johnson Method, (2) Orthogonal Method, and Wave Crest Method.

6.3.1 WAVE CREST METHOD

The first manual method that was utilized for the construction of refraction diagrams is known as the wave crest method (Johnson et al., 1948).

- A wave crest in deep water having the proper orientation is constructed using a hydrographic chart of the study area.
- A new crest position is drawn. This is accomplished by advancing points along the wave crest by an integral number of wavelengths and then a new position is drawn normally.
- The process is continued until the crest pattern from deep water to the shoreline is constructed.
- The advancing wavelength can be calculated given both the deep water wavelength and the local average water depth over which the wave will advance using the following equation

$$\frac{C}{C_o} = \frac{L}{L_o} = \tanh \frac{2\pi d}{L}. \tag{6.23}$$

A graphic template (Figure 6.9) that graphically solves the above equation has been constructed to aid in the plotting of crest positions (Wiegel, 1964).

- Orthogonal lines are constructed normal to the wave crests at desired intervals after the new crest position for the advancing wave train is drawn.

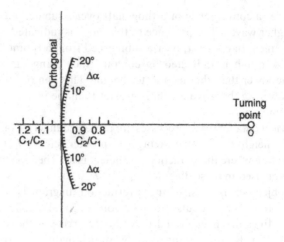

FIGURE 6.9 Template for wave crest method.

(From U.S. Army Coastal Engineering Research Center, 1977. Courtesy of the USA-COE)

6.3.2 ORTHOGONAL METHOD

A second manual graphical method based on Snell's Law for constructing refraction diagrams is known as the orthogonal method. Given refraction diagram is shown in Figure 6.10. Consider a train of waves propagating over a step where the water depth instantaneously decreases from d_1 to d_2. This causes the wave celerity and length to

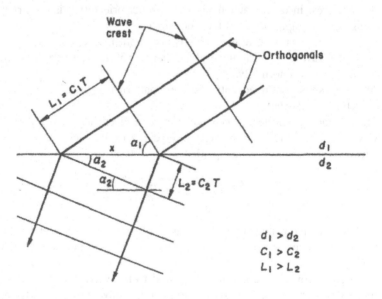

FIGURE 6.10 Diffraction using Snell's Law.

(From U.S. Army Coastal Engineering Research Center, 1977. Courtesy of the USA-COE)

decrease from C_1 and L_1 to C_2 and L_2, respectively. *For an orthogonal spacing x and a time interval T*, $\sin \alpha_1 = C_1 T/x$ *and* $\sin \alpha_2 = C_2 T/x$. *Dividing yields*

$$\frac{\sin\alpha_1}{\sin\alpha_2} = \frac{C_1}{C_2} = \frac{L_1}{L_2}, \qquad (6.24)$$

which is Snell's Law for wave refraction. Applying Equation (6.24) to wave refraction over a gradually varying bottom slope, α_1 and α_2 become the angles between the wave crest and bottom contour line at successive points along an orthogonal as a wave propagates forward, and L_1 and L_2 become the wavelengths at the points where α_1 and α_2 are measured. When waves propagate shoreward over bottom contours that are essentially straight and parallel as shown in Figure 6.11. Then the following equations can be written:

$$\frac{\sin\alpha_o}{L_o} = \frac{\sin\alpha}{L} = \frac{1}{x}. \qquad (6.25)$$

If we choose B_o and B so that the orthogonal lengths equal L_o and L as shown, then

$$K_r = \sqrt{\frac{B_o}{B}} = \sqrt{\frac{\cos\alpha_o}{\cos\alpha}}, \qquad (6.26)$$

where

$$\alpha = \sin^{-1}\left(\frac{C}{C_o}\sin\alpha_o\right). \qquad (6.27)$$

Equations (6.26) and (6.27) allow one to determine the refracted wave height and wave crest orientation or orthogonal direction when waves refract over essentially uniform nearshore hydrography. This process is as follows:

- Locate the depth contour represented by $d/L = 0.5$ on the hydrographic chart for the area of interest.
- Label each of the shallower contours in terms of the relative depth d/L_o.

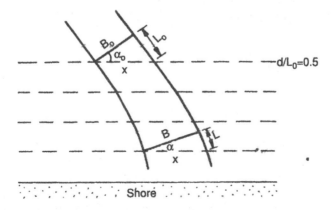

FIGURE 6.11   Wave propagation over parallel bottom contours.

- For each contour line and the one landward of it calculate the ratio of wave celerities C_1/C_2 where C_1 is the celerity at the deeper contour. From Equation (6.28):

$$\frac{C_1}{C_2} = \frac{\tanh\left(\dfrac{2\pi d_1}{L_1}\right)}{\tanh\left(\dfrac{2\pi d_2}{L_2}\right)}, \qquad (6.28)$$

where d/L can be determined by trial and error for Equation (6.29) rearranged as follows:

$$\frac{d}{L_o} = \frac{d}{L}\tanh\left(\frac{2\pi d}{L}\right). \qquad (6.29)$$

- Starting at the two most seaward contours, construct a line that is a mid-contour halfway between these two contours.
- Extend the incoming deep water orthogonal straight to the mid-contour, and construct a line tangent to the mid-contour at the intersection of the mid-contour and the incoming orthogonal.
- Lay the template (Figure 6.9) with the line marked orthogonal over the incoming orthogonal and $C_1/C_2 = 1.0$ at the intersection of the mid-contour and the orthogonal.
- Rotate the template around the turning point until the calculated value of C_1/C_2 intersects the tangent to the mid-contour line. The line of the template labeled orthogonal now lies in the direction of the outgoing orthogonal.
- With a pair of triangles, move the outgoing orthogonal to a parallel position such that the incoming and outgoing orthogonals connect and the lengths of these two orthogonal lines are equal.
- Repeat the above procedure at successive contour intervals to extend the orthogonal line forward toward the shore.
- Repeat the above procedure for additional orthogonal lines as needed.

The orthogonal spacing ratio b_o/b must be determined from a graphical refraction analysis as shown in Figure 6.8.

6.4 WAVE DIFFRACTION

Wave diffraction is concerned with the transfer of wave energy across wave rays. Refraction and diffraction take place simultaneously (Figure 6.11). Therefore, it is necessary at times to compute refraction and diffraction together. It is possible to define situations that are predominantly affected by refraction or by diffraction. Wave diffraction is specifically concerned with zero depth change and solves for sudden changes in wave conditions such as obstruction that cause wave energy to be forced across the wave rays.

6.4.1 REFRACTION/DIFFRACTION COMPUTER MODELS

A series of programs are available that deal with diffraction, in addition to modeling wave refraction and shoaling. An example of these computer models is RCPWAVE (Ebersole, 1985; Ebersole et al., 1986). This program is limited to open coast areas without structures or islands. REFDIF1 was developed for monochromatic wave refraction (Kirby and Dalrymple, 1991). The program REFDIF1 is a steady-state model based on a parabolic approximation solution to the mild slope.

6.5 WAVE REFLECTION

Wave energy will be approximately reflected from a wall when it hits one that is vertical, impermeable, and has a rigid surface (Figure 6.12). This reflected wave can be described as a standing wave (i.e. clapotis) as shown in Figure 6.13. A review of Figure 6.13 shows that there is no particle movement across antinode, and the surface profile does not change the nodes. As a consequence the water surface can be calculated as the sum of the incident wave height (η_i) and the reflected wave height (η_r) since $H_i = H_r$ (H_i – incident wave height, H_r – reflected wave height). Therefore the height of the water surface is given by Equation (6.36).

$$\eta = \eta_i + \eta_r = \frac{H\left[\cos\left(\frac{2\pi x}{L} - \frac{2\pi t}{T}\right) + \cos\left(\frac{2\pi x}{L} + \frac{2\pi t}{T}\right)\right]}{2}. \tag{6.30}$$

This equation can be reduced to the following:

$$\eta = H_i \cos\frac{2\pi x}{L} \cos\frac{2\pi t}{L}. \tag{6.31}$$

This equation represents the water surface for a standing wave which is periodic and has a maximum height of $2H_i$. In hydrodynamics, a clapotis is a non-breaking standing wave pattern caused, for example, by the reflection of a traveling surface wave train from a near vertical shoreline like a breakwater, seawall, or steep cliff.

The degree of wave reflection is defined by the reflection coefficient (C_r) which is defined as follows:

$$C_r = \frac{H_r}{H_i}, \tag{6.32}$$

where H_r is the reflected wave height, and H_i is the incident wave height.

$C_r > 1$ equals total reflection
$0 < C_r < 1$ equals partial reflection
$C_r = 0$ equals no reflection

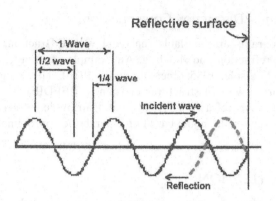

FIGURE 6.12   Wave reflection.

(From U.S. Army Coastal Engineering Research Center, 1984. Courtesy of the USA-COE)

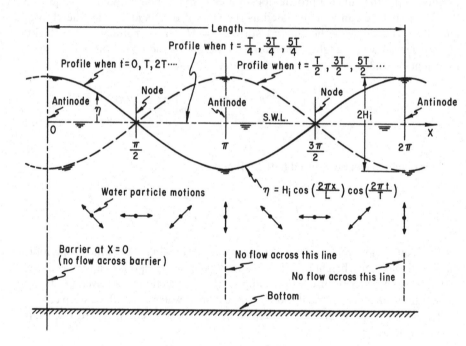

FIGURE 6.13   Standing waves system, perfect reflection from a vertical barrier.

(From U.S. Army Coastal Engineering Research Center, 1984. Courtesy of the USA-COE)

Water particle motions at nodes when $C_r = 1$ water particle motions are horizontal and all the wave energy is kinetic energy as shown in Figure 6.14. In contrast, at the antinodes, water particle motions are vertical and all of the wave energy is potential energy.

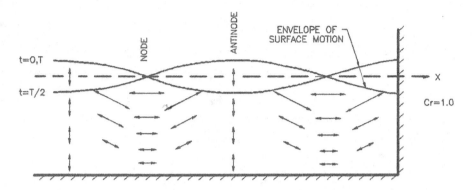

FIGURE 6.14   Visualization of water particle movement when $Cr = 1$.

(From U.S. Army Coastal Engineering Research Center, 1984. Courtesy of the USA-COE)

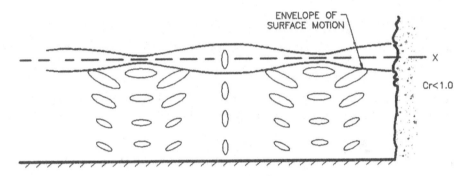

FIGURE 6.15   Visualization of water particle movement when $Cr < 1$.

(From U.S. Army Coastal Engineering Research Center, 1984. Courtesy of the USA-COE)

The water surface envelope develops when the reflection coefficient is less than 1 ($C_r < 1$) as shown in Figure 6.15. As the reflection coefficient decreases toward zero, the water surface profile and water particle path change toward the form of a normal progressive wave.

6.6   WAVE RUN-UP

6.6.1   INTRODUCTION

Wave run-up is the height (R) that it attains running up the shore before its energy is dissipated due to friction and gravity (Figure 6.16). Specifically, the wave run-up (R) is the combination of both wave set-up and swash uprush (Douglass, 1990; Douglass, 1992; Shih et al., 1994). The run-up needs to be added to the water level that was reached as a result of tides and wind set-up (Figure 6.16). The remaining energy will energize a bore after a wave breaks that will run up the face of a beach or sloped

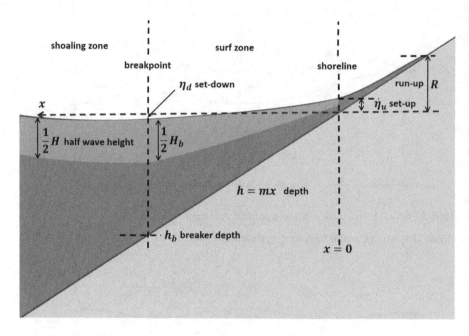

FIGURE 6.16 Wave set-down, wave set-up, and wave run-up.

(From Courtesy of the Wikipedia Creative Commons CC0 License. Available at: www. coastalwiki.org/w/index.php?title+Wave_run-up&oldid=79944)

shore structure. Waves are defined as wind waves, tsunamis, or seiches. The run-up is dependent on the following: (1) the incident deep water wave height and period, (2) the surface slope and profile form if not planar, (3) the depth d_s in front of the slope, and (4) the roughness and permeability of the slope face. Wave run-up has been approached by both model studies and analytical methods. In the following, each of these two approaches will be discussed.

6.6.2 Laboratory Model Studies

A typical laboratory wave run-up study with a monochromatic wave plot of experimental data is presented in Figure 6.17. These data are for an impermeable smooth planar slope with R/H_0' between 0.1 and 5. Similar plots for other slope conditions are available through the U.S. Army Coastal Engineering Research Center (1984). A review of Figure 6.17 indicates that, for a given structure slope, steeper waves (higher H_0'/gT^2) have a lower relative run-up $\left(R/H_0'\right)$. In addition, for most beach and revetment slopes the wave run-up increases as the slope becomes steeper. The effect of slope surface condition on wave run-up is presented in Table 6.2. The factor r is the ratio of the wave run-up on the given surface to that on a smooth impermeable surface. A selection of values for surfaces ranging from concrete to tetrapods is presented in Table 6.2. Laboratory experiments provided this information and gave an indication of the effect of slope surface condition on wave run-up. The factor r is the

TABLE 6.2
Run-Up Factors for Different Slope Conditions

Slope Facing	r
Concrete slabs	0.9
Placed basalt blocks	0.85–0.9
Grass	0.85–0.9
One layer of riprap on an impermeable base	0.8
Placed stones	0.75–0.8
Round stones	0.6–0.65
Dumped stones	0.5–0.6
Two or more layers of riprap	0.5
Tetrapods	0.5

Source: Adapted from Battjes (1974).

ratio of the run-up on the given surface to that on a smooth impermeable surface and may be multiplied by the run-up determined (Figure 6.17).

6.6.3 ANALYTICAL APPROACHES

For waves collapsing on the beach, the run-up magnitude can be given by the empirical formula of Hunt (1959).

$$R \approx \eta_u + H\xi, \tag{6.33}$$

where

$\eta_u \approx 0.2H$ is the wave set-up
H—the offshore significant wave height
ξ—surf similarity parameter

$$\xi = \frac{\tan\beta}{\sqrt{H/L}} = T \tan\beta \sqrt{\frac{g}{2\pi H}}, \tag{6.34}$$

where

$L = gT^2 / (2\pi)$ is the offshore wavelength
β—the beach slope
T—wave period

The horizontal wave incursion is approximately given by $R/\tan\beta$.

A number of additional empirical equations have been proposed for the run-up (Douglas, 1990; Van der Meer and Stam, 1992) refer to Table 6.3.

TABLE 6.3
Run-Up Formulas

Formula	Reference
$\dfrac{R}{H_o} = 1.84\xi rp$	Hunt (1959)
$\dfrac{R_2}{H_o} = 1.86\xi_o^{0.71}$	Mase and Iwagaki (1984)
$\dfrac{R_2}{H_o} = 0.83\xi_o + 0.2$	Holman (1986)
$R_2 = 0.099\sqrt{\dfrac{H_o}{1.4287}L_o}$	Nielsen and Hanslow (1991)
$\dfrac{R_2}{H_o} = \dfrac{0.12}{\sqrt{\dfrac{H_o}{L_o}}}$	Douglass (1992)
$\dfrac{R_2}{H_o} = 1.49\xi_o + 0.34$	Hedges and Mase (2004)
$R_2 = 1.1\left[0.35\beta_f\sqrt{H_oL_o} + 0.5\sqrt{H_oL_o\left(0.563\beta_f^2 + 0.004\right)}\right]$	Stockdon et al. (2006)
$\dfrac{R_2}{H_o} = C\left(\dfrac{h}{x_h}\right)^{\frac{2}{3}}$	Mather et al. (2011)
$\dfrac{R_2}{H_o} = 4\beta_f^{0.3}\xi_o$ and $R_2 = 4\beta_f^{1.3}\sqrt{H_oL_o}$	De La Pena et al. (2012)

Source: Adapted from Roux (2015).

Note: R—wave run-up; H_o—deep water wave height; ξ_o—Iribarren number; r—roughness factor; p—porosity factor; R_2—wave run-up exceeded by 2% of the record; L_o—deep water wavelength; β_f—beach face slope; X_h—hor. distance to closure depth; h—closure dept.

An equation for the run-up R_2 which is exceeded by only 2% of the waves has been presented by Stockdon et al. (2006) based on a large ξ data set.

$$R_2 = 1.1\left(\eta_u + 0.5\sqrt{S_W^2 + S_{ig}^2}\right),\ \xi \geq 0.3,\ R = 0.43\sqrt{HL},\ \xi < 0.3, \qquad (6.35)$$

where

$\eta = 0.35H\xi$—wave set-up,
$S_W = 0.75H\xi$ — swash uprush related to incident waves,
$S_{ig} = 0.06\sqrt{HL}$ — additional uprush related to infragravity waves.

Infragravity waves are ocean surface wind waves with a period ranging from 25 to 250 s. They are generally small on the open ocean but close to the coast they can be up to a few meters in height thus dominating water motion during storms.

The factor 1.1 in Equation 6.35 takes into account the non-Gaussian distribution of run-up events. For steep reflective beaches (m>0.1) the run-up increases with increasing beach slope (approximately linear dependence), while for gently sloping dissipative beaches (m<0.1) the dependence on beach slope is weak or absent (Gomes da Silva et al., 2020). In these cases, run-up is dominated by infragavity waves. These types of waves yield a small run-up that increases with increasing wave height (approximately linear). A dissipative coast is a gently sloping flat beach with a broad surf zone and typically sand. In contrast, a reflective coast is a steep beach with a berm and a narrow surf zone. It typically is composed of coarse sand and gravel. In addition, field observations and numerical models indicate a dependence of intra-gravity swash on the frequency spread and the directional spread of incident waves. The largest infragavity swash occurs for incident waves with a small directional spread and a large frequency spread.

The run-up R_2 is exceeded by only 2% of the waves if the backshore berm or foredune of dissipative sandy beaches is artificially protected with gravel or cobbles. This occurs when the water level during storms reaches higher than the toe of the cobble berm revetment. The R_2 can be estimated from the approximate empirical formula (Blenkinsopp et al., 2022).

$$R_2 \approx 4.14 \; h_{toe} m_{berm} + 0.66, \qquad (6.36)$$

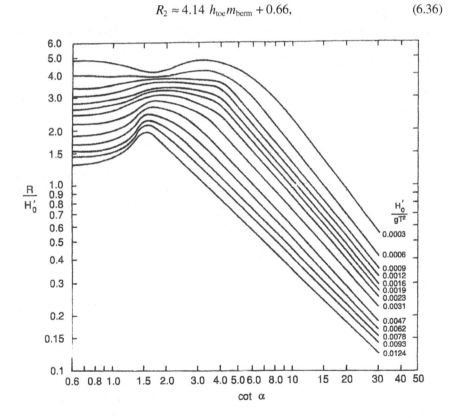

FIGURE 6.17 Dimensionless run-up on smooth impermeable slopes versus bottom slope and incident deep water wave steepness; $1 < ds/H'_o < 3$.

(From U.S. Army Coastal Engineering Research Center, 1984)

where

h_{toe} is the water depth at the toe of the berm
m_{berm} is the berm slope

The general applicability of empirical formulas of run-up based on simple representations of beach and shoreface is limited due to the influence of the more detailed
characteristics of the local shoreface bathymetry. This is similar to the limited applicability of empirical formulas for the wave set-up, which is a substantial component
of the run-up. Accurate estimates of the wave run-up require in-situ observations or
detailed numerical models.

6.7 SUMMARY

When a wave encounters a coastal structure or beach a process begins to dissipate
energy. The effects of waves on the coastal zone can be divided into three zones
which are (1) offshore zone, (2) shoaling zone, and (3) surf zone.

The surf zone has been described by Barua (2016) as the shoreline region that
exists from the initial wave breaking seaward to the still water level shoreward. This
zone shifts continuously from shoreward during high tide to seaward during low tide.
In addition, the zone shifts shoreward during low waves and then to seaward during
high waves.

The beach configuration changes with the amount of wave energy being dissipated. This configuration of the beach changes is reflected between winter storm
waves and the summer more moderate wave climate. The effectiveness of the
waves to mobilize sediment depends in large part on whether it can feel the bottom.
That is the depth that movement of water particles due to waves extend. In deep
water this depth (d) is related to the wavelength ($d = L/2$). This depth is called the
wave base. As the wave starts undergoing shoaling the water particles start to feel
the bottom. This near-shore process of surface gravity waves decreasing celerity is
referred to as wave shoaling. When water waves approach a coastline their water
particle orbits undergo a change from circular to elliptical. At this point the wave
height increases and wave length decreases as the depth decreases.

Wave shoaling, refraction, and diffraction transform the waves from deep water
to the point where they break. The corresponding wave height begins to decrease
markedly, because of energy dissipation. The sudden decrease in the wave height
is used to define the breaking point and determine the breaking parameters. One of
these parameters is the wave steepness (H/L) which is defined as the ratio of wave
height (H) divided by wavelength (L). When the wave steepness becomes ($H/L \geq 1/7$),
it becomes unstable and breaks (i.e. surf zone). Four types of breaking waves occur
in the surf zone on the coast. These breaking waves are spilling, plunging, collapsing, and surging. The type of breaking wave depends on the Iribarren number (ξ_i) or
surf similarity parameter. The surf similarity parameter compares the wave surface
slope to the bed slope (m), the still water depth h, the significant incident water height
H_s, and the wave period T. There are a number of issues involved when waves travel
from deep into shallow water (i.e. shoaling). These issues are shoaling, refraction,

and diffraction. Wave transformation describes the changes that occur in waves as they move from deep into shallow water. Waves rarely approach shore at a perfect 90-degree angle. As waves approach the shoreline, they bend so that the wave crests become nearly parallel to the shore. Different segments of the wave crest will travel at different speeds since wave speed in shallow water is proportional to the depth of water. This change in speed of the wave will cause refraction.

Wave run-up is the height (R) that it attains running up the shore before its energy is dissipated due to friction and gravity. Specifically the wave run-up (R)is the combination of both wave set-up and swash uprush. The run-up needs to be added to the water level that was reached as a result of tides and wind set-up. The remaining energy will energize a bore after a wave breaks that will run up the face of a beach or sloped shore structure.

REFERENCES

Barua, D. K. (2016). The surf zone. Wide Canvas- *Science and Technology*, Nov, 3.

Battjes, J. A. (1974). Surf Similarity, Proceedings of 14th Coastal Engineering Conference, ASCE: pp. 466–480.

Blenkinsopp, C. E., Bayle, P. M., Martin, K., Foss, O. W., Almeida, L. P., Kaminsky, G. M., Schimmels, S., and Matsumoto, H. (2022). Wave run-up on composite beaches and dynamic cobble berm revetments. *Coastal Engineering*, 176, 1041–1148.

De la Pena, J. M., Sanchez Gonzalez, J. F., Diaz-Sanchez, R., and Martin Huescar, M. (2012). Physical model and revision of theoretical run up. *Journal of Waterway, Port, Coastal and Ocean Engineering*, 140(4), 1–9.

Douglass, S. L. (1990). Estimating runup on beaches: A review of the state of the art: C.R. CERC-90-3, U.S. Army Engr. Waterways Experiment Station, Vicksburg, MS.

Douglass, S. L. (1992). Estimating extreme values of run-up on beaches. *Journal of Waterways, Port, Coastal and Ocean Engineering*, 118, 220–224.

Ebersole, B. A. (1985). Refraction-diffraction model for linear waves. *Journal of Waterway, Port, Coastal, and Ocean Engineering*, III(WW6), 939–953.

Ebersole, B. A., Cialone, M. A., and Prater, M. D. (1986). RCPWAVE—A linear wave propagation model for engineering use. Technical Report CERC – 86-4. U.S. Army Engineer Waterways Experimental Station, Vicksburg, MS.

French, P. W. (1997). *Coastal and Estuarine Management*, Routledge Environmental Management Series. London: Routledge.

Galvin Jr., C. J. (1968). Breaker type classification on three laboratory beaches. *Journal of Geophysical Reserch*, 73(12), 3651–3659.

Gomes da Silva, P., Coco, G., Garnier, R., and Klein, A. H. F. (2020). On the prediction of runup, setup, and swash on beaches. *Earth-Science Reviews*, 204, 103148.

Hedges, T. S., and Mase, H. (2004). Modified hunt equation incorporating wave setup. *Journal of Waterway, Port, Coastal and Ocean Engineering*, 130, 109–113.

Holman, R. A. (1986). Extreme value statistics for wave runup on a natural beach. *Coastal Engineering*, 9(6), 527–544.

Hunt, I. A. (1959). Design of seawalls and breakwaters. *Journal of Waterways and Harbors Division*, 85, 123–152.

Johnson, J. W., O'Brien, M. P., and Isaacs, J. D. (1948). Graphical construction of wave refraction diagrams. U.S. Navy Hydrological Office Publication, No. 605.

Kinsman, B. (1965). *Wind Waves: Their Generation and Propagation on the Ocean Surface.* Englewood Cliffs, NJ: Prentice-Hall.

Kirby, J. T., and Dalrymple, R. A. (1991). User's manual, combined refraction/diffraction model, Ref-Dif 1. Version 2.3. Center for Applied Coastal Research, Department of Civil Engineering. University of Delaware. Newark, DE.

Mase, H., and Iwagaki, Y. (1984). Run-up of random waves on gentle slopes. *Coastal Engineering Proceedings*, 1(19), 40.

Mather, A., Stretch, D., and Garland, G. (2011). Predicting extreme wave run-up on natural beaches for coastal planning and management. *Coastal Engineering Journal*, 53(2), 87–109.

Nielsen, P., and Hanslow, D. J. (1991). Wave runup distributions on natural beaches. *Journal of Coastal Research*, 7(4), 1139–1152.

Okazaki, S., and Sunamura, T. (1991). Re-examination of breaker-type classification of uniformly inclined laboratory beaches. *Journal of Coastal Research*, 7(2), 559–564.

Roux, A. P. (2015, March). A re-assessment of wave run up formulae. Thesis submitted for Masters of Engineering, University of Stellenbosch.

Shih, S.-M., Komar, P. D., Tillotson, K. J., McDougal, W. G., and Ruggiero, P. (1994). Wave Run-up and sea-cliff erosion, Coastal Engineering, ASCE, Chapter 157: pp. 2170–2184.

Stockdon, H. F., Holman, R. A., Howd, P. A., and Sallenger, A. H. (2006). Empirical parameterization of runup, setup, swash, and runup. *Coastal Engineering*, 53, 573–588.

U.S. Army Coastal Engineering Research Centre. (1977). *Shore Protection Manual, Vol. 1.* Department of the Army, Corps of Engineers, Washington, DC.

Van der Meer, J. M., and Stam, C.-J. M. (1992). Wave runup on smooth and rock slopes of coastal structures. *Journal of Waterway, Port, Coastal and Ocean Engineering, American Society of Civil Engineers*, 118, 534–550.

Wiegel, R. L. (1964). *Oceanographical Engineering.* New Jersey: Prentice-Hall, p. 532.

Part III

Coastal Soil and Rock Materials

7 Sources and Characteristics of Beach Material

7.1 INTRODUCTION

Soil is any unconsolidated material composed of discrete solid particles and inter-stitial gas and/or pore liquids (Sowers and Sowers, 1961). As products of various geological processes, soils are extremely variable mixtures of weathered minerals (i.e., naturally occurring, solid, chemical elements, or compounds formed by a geological process). These minerals are the result of fragmented rocks (i.e., aggregates of minerals or natural glasses), highly variable in composition, grain size, and spatial arrangement of their particles, which may contain small amounts of organic matter, mostly decaying plant debris and animal residues. The pore fluids may chemically affect the nature of the soil. This effect is especially predominate if the soil is fine-grained and depends on its constituents, such as dissolved salts and organic compounds. Soils can be either dry, partially saturated, or saturated. Terrestrial soils are usually partially saturated. In contrast, marine soils or sediments are usually totally saturated (i.e., the interparticle voids are completely filled with seawater) if they are deposited under water. Beach sediments may exist in all three states. In some cases, the interparticle voids can contain gases, mostly produced by biogenic degradation of organic matter or introduced through geochemical processes. The high variability of soils is reflected by the wide range of their geophysical, geochemical, and mechanical properties.

Soils are conveniently sorted, or classified, into groups showing similar behavior or based on their significant properties. Soil classification has proved to be a valuable tool for the characterization and correlation of soils, and the efficient use and interpretation of both geological and engineering information. In general, soils (i.e., sediments) can be broken into two primary categories depending on whether they were transported into or were formed within the environment: (1) allochthonous and (2) autochthonous (Figure 7.1).

7.1.1 SEDIMENT SOURCES

Littoral deposit distributions and dimensions in the Cascadia margin are controlled principally by a combination of tectonics, sediment supply, and the wave climate. The primary boundaries to longshore transport are geologic structures formed from a combination of either accreted terranes and/or deformation in this active sub-duction zone. The size and location of sediment deposits in the coastal zone are

DOI: 10.1201/9781003454212-10

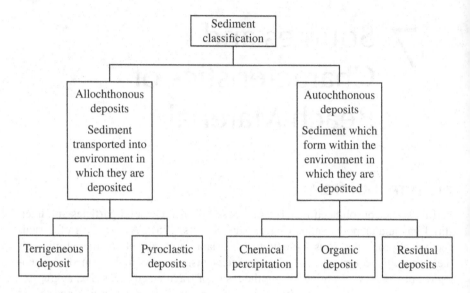

FIGURE 7.1   Basic divisions of sediments.

(Chaney and Almagor, 2016. Republished with permission of Taylor & Francis)

dependent of whether the coast is either uplifted or subject to subsidence relative to present sea level. This process determines the composition and amount of sediment supplied to beaches by sea cliff retreat. The source of the regional variability of littoral sand accumulation corresponds to the spatial distribution of major river sources. The rivers providing sources of littoral sand include the Columbia, Eel, Klamath, and Umpqua Rivers.

The stability of a coastline depends on the amounts of sand supplied annually to or lost from the coast. The primary sources of this sand material are the following:

1. Major streams
2. Small streams or gullies
3. Cliff erosion and slides
4. On shore movement of sand by wave action
5. Wind action

The sediment that is carried into the ocean from the land ranges in size from clay material (<0.007 mm) to large sand and rock fractions. Sediment transport by streams can be considered in two parts: wash load and bed load. Wash load is that part of the total sediment load composed of particles finer than a limiting size, which is normally washed into and through a river reach. In contrast, bed load is the soil, rock particles or other debris rolled along the bottom of a stream by the moving water.

Sediment sizes existing on beaches exposed to wave action have very little particles finer than <0.002 mm. Finer material is usually carried seaward into deep water as a result of wave action and currents in the nearshore area. Therefore material with sizes >0.002 mm (silt sizes) will be found on the beach. The size of material present on a particular beach is controlled by wave energy.

7.2 COMPOSITIONAL PROPERTIES

In coastal engineering work, knowledge of the sediment composition is not usually important in its own right, but it is closely related to other important parameters such as sediment density, erosional resistance, and fall velocity.

7.2.1 MINERALS

The mineral most commonly found (approximately 70%) in beach sand is quartz. This is because of quartz's resistance to both physical and chemical changes and its occurrence in terrestrial rocks. On beaches in the temperate latitude, more than 90% of the materials are quartz and feldspar (Krumbein and Sloss, 1963). On individual beaches, the percentage of quartz or other mineral can range from 0% to 100%. Crustal rocks are comprised of approximately 12% quartz. In contrast, approximately 50% of crustal rocks are comprised of feldspars (Ritter, 1986). The difference in percentages that remain in beach sands is that feldspars are more subject to chemical weathering. The weathering process converts the feldspars to clay minerals, quartz, and solutions. Since quartz is so inert, it accumulates during the weathering processes. Coastal sediments close to sources of igneous and metamorphic rocks contain feldspars and silicates. These rocks typically occur in mountainous and glaciated coasts. On shorelines that are at a distance from mountains the percentage of quartz sand and clay increase. This is due to weathering having had more time to reduce the relative proportions of feldspars and related silicates.

Carbonate sands owe their formation to organisms that precipitate calcium carbonate. Calcium carbonate can be deposited as calcite or aragonite. Aragonite is unstable and can change into either a solid calcite or aqueous solution. Calcite (i.e., limestone), can be altered to dolomite by the partial replacement of calcium with magnesium. Carbonate sands can constitute up to 100% of the beach material. This is particularly the case where it is produced in the local marine environment and there are limited terrestrial sediment sources. These types of sands are generally composed of a combination of the following: shell and shell fragments, oolites, coral fragments, and algae fragments (Halimeda, foraminiferans, etc.) (Chaney et al., 1982).

Heavy minerals typically form a small percentage of beach sands. Minerals are described as heavy because their specific gravities are >2.87. The color of these minerals is typically black or reddish. These heavy minerals in large concentrations may color the entire beach. These minerals are the following: andalusite, apatite, augite, biotite, chlorite, diopside, epidote, garnet, hornblende, hypersthene-enstatite, ilmenite, kyanite, leucoxene, magnetite, muscovite, rutile, sphene, staurolite, tourmaline, zircon, and zoisite (Pettijohn, 1957). The amount of these minerals in beach sands is a function of both (1) their distribution in the source rocks of the littoral sediments and (2) the weathering process.

7.2.2 Density and Unit Weight

Density is defined as the mass per unit volume of a material. It is measured in SI units as kilograms per cubic meter (kg/m^3). This relationship is shown in Equation (7.1).

$$\rho = \frac{m}{V},\tag{7.1}$$

where:

    ρ—density
    m—mass
    V—volume

Unit weight in turn is defined as follows:

$$\gamma = \rho g,\tag{7.2}$$

where:

    g—gravity
    γ—unit weight

The density of a sediment is a function of its composition. Minerals commonly encountered along the coast include quartz, feldspar, clay minerals, and carbonates. Densities for these minerals are given in Table 7.1.

Quartz sand is composed of the mineral silicon dioxide. In contrast, feldspar refers to a closely related group of metal aluminum silicate minerals. Common clay minerals are illite, montmorillonite, and kaolinite.

Common carbonate minerals include calcite, aragonite, and dolomite. However, carbonate sands are usually not simple, dense solids; but rather the complex products of organisms which produce gaps, pores, and holes within the structure, all of which tend to lower the effective density of carbonate sand grains. Thus, carbonate sands frequently have densities less than quartz.

The density of a sediment sample may be calculated by adding a known weight of dry sediment to a known volume of water. The change in volume is measured; this is the volume of the sediment. The sediment mass (= weight/acceleration of gravity) divided by its volume is the density. A complicating factor is that small pockets of air will stay in the pores and cling to the surfaces of almost all sediments. To obtain an accurate volume reading, this air must be removed by applying a strong vacuum over the sand-water mixture.

Sediments found along the shoreline are typically rock composed of several minerals. The densities of rocks commonly found along the shoreline are listed in Table 7.1. These rocks are also used for riprap, which are large blocks or boulders. Lines 5 and 6 of Table 7.2 deal with carbonate rocks, the dolomites and limestones.

TABLE 7.1
Densities of Common Coastal Sediments

Mineral	Density (kg/m³)
Quartz	2.648
Feldspar	2.560–2.650
Illite	2.660
Montmorillonite	2.608
Kaolinite	2.594
Calcite	2.716
Aragonite	2.931
Dolomite	2.866

Source: U.S. Army Corps of Engineers, *Coastal Engineering Manual* (2008). Courtesy of U.S. Army.

TABLE 7.2
Average Densities of Rocks Commonly Encountered in Coastal Engineering

Number	Rock Type	Mass Density (kg/m³)
1	Basalt	2.740
2	Dolerite-Diabase	2.890
3	Granite	2.660
4	Sandstone	2.220
5	Dolomite	2.770
6	Limestone	2.540

Source: U.S. Army Corps of Engineers, *Coastal Engineering Manual* (2008). Courtesy of U.S. Army.

7.2.3 Specific Weight and Specific Gravity

The specific weight of a material is its density times the acceleration of gravity g. The acceleration of gravity is 9.807 m/s², and thus, specific weight is measured in kilograms (kg) per meter (m) squared per second(s) squared (kg/(m² * s²)) in SI units. The specific gravity of a material is its density divided by the density of water at 40°C, which is 1000.0 kg/m³. Specific gravity is a dimensionless quantity.

7.2.4 Strength

The maximum stress which the particle can resist without failing is its strength. To measure strength, Pascal is used as the SI unit of stress. The Pascal multiple is 1 million Pascals, or 1 mega Pascal (MPa). Strength can then be determined using a variety of test such as unconfined ultimate strength in compression, which is equivalent to crushing strength. Tensile strength, which is significantly less than compressive strength but is usually proportional to compressive strength.

The most common material found along the shore is the quartz sand grain. A single crystal of quartz has a typical strength of 2500 MPa. In contrast, a sandstone, which is a composite of many sand grains, is surprisingly weak with strength typically less than 100 MPa, or less than 4% of the strength of the single crystal. For this reason, sandstone is rarely used in coastal engineering construction.

The difference in strength between a quartz crystal and composite sandstone is due to weak intergranular cement and flaws. These flaws typically are the following: grain boundaries, bedding planes, cleavage, and joints that have a higher probability of being present in larger pieces (Handin, 1966).

For calcium carbonate, in contrast to quartz strength single crystals of calcite are weak (~14 MPa, depending on orientation) compared to single crystals of quartz (~2500 MPa). Limestone rock, made from interlocking calcium carbonate crystals, is much stronger than single crystals of calcium carbonate. In addition, limestone rock is somewhat stronger than sandstones.

Where available, rocks classified commercially as trap rock (dense basalt, diorite, and related rocks) or granite (including rhyolite and dense gneiss) make even better riprap, with strength typically on the order of 140 to 200 MPa.

A typical specification for rock used as riprap in coastal engineering, extracted from a natural deposit is as follows: The stone shall be free of cracks, seams, and other defects that would tend to increase unduly its deterioration from natural causes or breakage in handling or dumping. The stone shall weigh, when dry, not less than 2400 kg/m^3 (150 pounds per cubic foot). The inclusion of objectionable quantities of sand, dirt, clay, and rock fines will not be permitted. Selected granite and quartzite, rhyolite, traprock, and certain dolomitic limestones generally meet the requirements of these specifications.

7.3 TEXTURE, STRUCTURE, AND CHEMICAL COMPOSITION OF SOIL PARTICLES

7.3.1 INTRODUCTION

The properties of a soil are a function of both its texture and structure, and the chemical and mineralogical composition of its particles. Some of these properties are its capacity to adsorb water, its plasticity and cohesion, and its changes in mechanical behavior due to variations in environmental conditions and time, such as its compressibility, permeability, strength, and stress transmission. The sediments together with fossils can also provide data needed for the reconstruction of the geological processes that have formed the soil.

Texture is concerned with the size of the soil's minerals and rock particles, specifically referring to the relative proportions of particles of various sizes. The distribution and orientation of the particles within the soil mass, and their arrangement into groups or aggregates is termed structure. These arrangements are seen in hand specimens and soil sections, or by ultra-thin methods. Large features of sedimentary origin, such as stratification, bedding, lamination, ripple and wave marks, uncon-formities, and deformations of various origins, define sedimentary structures. Composition refers to the nature and arrangement of the atoms in a soil particle.

The texture, structure, and composition, and consequently the physical properties, of detrital terrigenous and volcanic (lithogenous) sediments, and of sea-born biogenic (biogenous) and chemical (hydrogenous) sediments are very different. The particles of terrigenous and volcanic sediments produced by physical and chemical weath-ering processes are silicate and aluminosilicate minerals. These particles (except glacial rock floor and those recently introduced by submarine volcanic eruptions) are very resistant to both physical and chemical weathering. The size distribution of these sediments is not subject to ready change, and they remain unconsolidated and chemically unaltered in their depositional environments, even if they become subaerially exposed during sea level falls. By contrast, the biogenous and hydrog-enous sediments, which are mostly from shallow water environments, are subject to intensive diagenesis. These sediments commonly become cemented on the seafloor, and during exposure they readily alter chemically and mineralogically into typically either carbonates or evaporites. Such changes greatly affect the texture, structure, and composition, and consequently the physical properties of sediments.

7.4 SIZE, SHAPE, AND STRUCTURE

7.4.1 INTRODUCTION

Disintegration caused by impact, abrasion, grinding, and transportation of the sedi-ment solution to the sea produce discrete soil particles widely ranging in size, shape, and texture.

Soil particles vary in size from 1 nm ($1 \times 10E{-}8$ m $= 10$ Å) up to large rocks several meters in diameter, a range of one to more than a billion. Soil particles are conveniently separated into groups according to size, their ranges arbitrarily set, and each assigned a name. There are no universally accepted definitions of size fractions, and different size classifications are developed for different uses (e.g., geological, pedological, and agricultural soil engineering).

The degree of similarity in size of the particles in sediment is termed sorting. Engineers consider a soil poorly sorted if it is predominately of one particle size. By contrast, geologists consider a soil well sorted if it is predominately of one particle size. In this chapter the engineering interpretation is employed. The particle sizes, determined by mechanical size analyses, are plotted as a summation or frequency curve, the form of which give more information than can be readily visualized from a table of percentages, for example, the degree of sorting or grading. The very coarse particles, from gravel upward, are fragmented rocks and are usually highly variable in shape. They may be irregularly shaped, more or less rounded, or even flat. Fragmented rocks are usually a minor constituent and rated separately in soil because of their size.

The sand-, silt-, and clay-sized components of a soil may be of diverse origins and compositions. Although the shape of the individual particles may vary in a given soil, they are basically the same in each of these major size groups. Usually, the shape and size of the particles refers to the composition and the crystal structure of the grains. Sands and silts tend to be bulky, and are approximately equidimensional, reflecting the physical weathering process experienced during their transportation. Clays, the result of chemical weathering, typically possess platey, sheetlike shapes reflecting their crystal structures. The sphericity, or the degree of closeness to a sphere, is a measure of the grain shape. The degree of roundness (or that of angularity) refers to the sharpness of the edges and corners of a particle in the silt and larger size range. High degree of angularity of particles is indicative of recent fragmentation of a larger rock fragment, short transportation, or high resistivity to weathering, whereas high degree of roundness reflects the opposite. Roundness and sorting are normally closely related. Well-sorted deposits of coarse, well-rounded grains are typical products of winnowing and abrasion encountered in high-energy environments, such as beaches; while deposits in low-energy environments are likely consist of more poorly sorted material, with a fine-grained size and more angular particles. The degree to which a deposit is both well sorted (i.e., contains a variety of different particle sizes) and rounded defines its textural maturity (Folk, 1951). Surface texture refers to minor features of a surface of usually hard, sand-sized particles caused by transport process: rough surfaces (result of sediment fragmentation); smooth, polished, and shiny surfaces (abrasion); frosty and pitted (grain impact during eolian transportation); etched surfaces (dissolution); and striated (ice grinding).

7.4.2 PACKING OF COARSE-GRAINED/NONCOHESIVE SEDIMENTS

The grain skeleton of gravels, sands, and silts typically consists of particles in point-to-point contact with those surrounding them. A very wide range of grain particle arrangements (packing) can form, depending on the relative positions of the soil particles.

Packing is defined as the spacing or density pattern of mineral grains. Generally, the smaller the range of particle size (i.e., the more nearly uniform the particle size, poorly graded), the smaller the particles and the more angular their shapes, the greater their opportunity for constructing a loose structure with low density. By contrast, the poorer the particle size sorting (e.g., well graded), the denser the packing. This is caused by the smaller particles filling the voids between the larger particles. Table 7.1 presents the typical ranges of void ratios, porosities, and dry unit weights of a variety of soil types.

Particles having low sphericity, such as flat gravels and flaky micas, often construct arches within the soil that bridge over large voids, resulting in loose, yet tightly wedged, stable arrangements. These particles also may tightly interlock and form dense packing configurations. Flaky particles often form oriented structures, developed during sedimentation or resulting from movement by shear stresses or high pressure.

Both loose and dense soil structures are capable of supporting considerable static over-burden loads with little or no distortion. However, loose soils with grain sizes in the coarse, silt and fine sand range are inherently metastable as shown by Terzaghi (1925). Shocks and vibrations readily cause movement of the particles to denser, more stable arrangements.

Highly metastable, loose sediments are often produced by rapid deposition of silt and fine sand in quiet environments, such as beaches in low-energy environments and delta fronts of large rivers. Many of the fine silt-sized particles are actually land-derived minuscule mudstone particles, while others of nonclayey compositions have an adhering film of clay. Therefore silts can possess some clayey properties, such as plasticity, cohesion, and adsorption. The structure of these sediments is sustained by cohesive forces at the silt grain contact and the friction at the contact of the sand grains directly perched on one another (Figure 7.4). This structure readily collapses upon shaking, which causes complete loss of their grain-to-grain contacts (liquefaction), followed by settlement of the suspended grains, displacement of their pore water, and packing to a higher density. The porosity does not indicate the degree to which a soil is loose or dense. Porosity is defined as the ratio of the volume of the voids in a given mass of soil to its total volume.

The degree of compaction, termed relative density D_r, of a cohesionless soil can be quantitatively obtained by comparing the porosity of the soil with its loosest and densest possible states. Relative density is calculated using Equation (7.3). The test to determine the maximum unit weight of the soil involves some form of vibration, and that to determine the minimum unit weight requires pouring of oven-dried soil loosely into a container of known volume,

$$D_r = \frac{e_{max} - e}{e_{max} - e_{min}} \times 100\% = \frac{\gamma_{max}\left(\gamma - \gamma_{min}\right)}{\gamma_{max} - \gamma_{min}} \times 100\%, \tag{7.3}$$

where:

e is defined as the void ratio (ratio of the volume of the voids in a given soil to the volume of its solids).
γ is defined as the dry unit weight (weight per unit volume).
max and min denote densest and loosest conditions, respectively.

For uniformly sized particles, the void ratio typically ranges from $[e_{max} =]$ 92% to $[e_{min} =]$ 35% (Table 7.1). Use of relative density has been restricted to granular soils because of the difficulty of determining the minimum unit weight in clayey soils.

7.5 FINE-GRAINED SEDIMENTS

7.5.1 INTRODUCTION

In the case of clay there are strong correlations among the three classification groups. A clay particle is a mineral whose molecules are arranged in sheets that feature orderly arrays of silicon, oxygen, aluminum, and other elements (Lambe and Whitman, 1969). Clay particles are small and either platey or tubular. These particles are small because they originate from the chemical modification and disintegration of pre-existing mineral grains. In addition, the predominantly sheet-like minerals are not strong enough to persist in large pieces. The geologist's size classification defines a particle as clay if it is less than 0.0039 mm. In contrast, the engineer defines

TABLE 7.3

Relations Among Three Classifications for Two Types of Sediment

| Name of Sediment | Bases of Classification | | |
	Usual Composition	Size Range, Wentworth	Bulk Properties
Clay	Clay minerals (sheets of silicates)	Less than 0.0039 mm	Cohesive Plastic under stress Slippery Impermeable
Sand	Quartz (SiO_3)	Between 0.0625 and 2.0 mm	Noncohesive Rigid under stress Gritty Permeable

Source: U.S. Army Corps of Engineers, *Shore Protection Manual* (2008). Courtesy of U.S. Army.

a clay as particles less than 0.002 mm. Because clay particles are so small, it has a large surface area compared to its volume. Because of this large surface area chemical forces predominate over body forces (i.e., gravity). Thus surface and end areas of the clay particle are chemically active. When the clay particles are hydrated in its bulk form a characteristic cohesive, plastic, and slippery material occurs. Thus, the three classifications each identify the same material when describing "clay."

Several differences between clay and sand are summarized on Table 7.3. More inclusive discussions of sediment sizes, compositions, and bulk properties are given later in this chapter.

7.5.2 PORE FLUIDS AND CLAY MICROSTRUCTURE

Unlike silt and larger size particles, the tiny, sheet-like clay particles tend to stay in suspension within the water column for a long time. Physiochemical forces between the clay particles, and between them and the seawater, are responsible for this dispersion. The soil structure that forms on the coastal and beach sediments is dependent on the concentration of particles of this dispersion during their gravitational settling. Knowledge of the physiochemical properties and interactions between clay particles and the resultant microstructure of their deposits is fundamental to the understanding of soil behavior. Numerous investigations of the physiochemical interactions in clay systems have been conducted (e.g., Grim, 1968; Grimshaw, 1971; Moon, 1972; Mysels, 1959; van Olphen, 1977; Yariv and Cross, 1979). Bennett and Hulbert (1986) presented an excellent review of the historical development of clay physiochemistry and the state-of-the-art of clay microstructure. Modern electron microscopic techniques and methodologies have allowed direct observations of the microstructure of a wide range of submarine sediments (Bennett et al., 1990). The study of clay microstructure in recent sediments has found no correlation between microscale and macroscale of soil structures and the genetic interpretation of various morphologies revealed by electron microscopy.

7.5.3 Physiochemical Interactions

Physiochemical interactions are essentially electrical in character. Enumerated in a roughly decreasing order of strength they include repulsive forces (electrical repulsion between like-charged particles), covalent bonds (attraction between atoms resulting from the sharing of pairs of electrons), electrostatic interactions (attraction of unlike charges and repulsion of like charges as in ionic bonds), hydrogen bonds (attraction of hydrogen ions by electron-rich ions), and van der Waals forces (attraction between atoms or molecules due to their electrical asymmetry).

The clay minerals are chemically stable due to the strong, partially covalent bonding between the tetrahedral silicon units and the octahedral aluminum- and magnesium-hydroxyl units within the clay sheets (Table 7.2). These sheets are effected by intensive surface physiochemical interactions. Every clay particle carries an electrical charge, usually negative, arising from a number of sources. These sources are the following: the inexact charge balance on its outer faces (i.e., termination of the crystal lattice); isomorphous substitution and the random absence of cations in the crystal lattice; and to a lesser degree, from ionization (dissociation) of surface groups (mostly hydroxyls), adsorption of anions from the surrounding solution, and the presence of organic matter.

The smaller the particle the larger its specific surface and, consequently, its surface chemical activity (Figure 7.2). The specific surface is defined as the surface area of particle divided by either its mass or volume. The behavior of clay particles, in the range from approximately 1 mm to 1 nm in size termed colloids, is mostly determined by surface-induced electrical forces (physiochemical interactions: flocculation and dispersion) rather than by mass-derived forces (gravity: settling). In contrast to clays, silts and sands with specific surfaces at least 10,000 times larger are chemically inactive. Clay particles carry a net negative charge on their surfaces. This is a result of both isomorphous substitution and a break in the continuity of the structure at its edges.

TABLE 7.4
Properties of Major Types of Silicate Minerals

Property	Types of Clay		
	Montmorillonite	Illite	Kaolinite
Size (m)	0.01–1.0	0.1–2.0	0.1–3.0
Shape	Irregular flakes	Irregular flakes	Hexagonal crystals
Specific surface (m²/kg)	700–800	100–200	5–20
External surface	High	Medium	Low
Cohesion, plasticity	High	Medium	Low
Cation exchange capacity (mEq/100 g)	80–100	15–40	3–15

Source: U.S. Army Corps of Engineers, *Shore Protection Manual* (2008). Courtesy of U.S. Army.

The clay particles to achieve electroneutrality attract positively charged ion or cations and polar water molecules from the surrounding solution onto their surfaces. This process is called hydration. Hydration occurs either directly onto the particle surface or indirectly, around the adsorbed cations, resulting in their phenomenal volume growth (sevenfold or more). Due to their size and the repulsion between the surface-attracted cations and water molecules, the hydrated ions move away from the particle surface to equilibrium positions where they best satisfy both surface attraction and cation repulsion forces between individual particles and their concentration decreasing exponentially with distance from the particle surface (Figure 7.3). The curve shapes depend on the mineralogical nature and chemical composition of the particles.

The surface-held water is denser than the water of the surrounding solution, forming double layers (approximately 400 Å thick) around the clay particles. The water nearest to the surface of the clay particle surface (approximately 5 Å) is strongly attracted to it, gradually becoming free away from it. At the margins of the double layers, the concentration of cations and water molecules becomes equal to that of the surrounding solution.

The cations are generally weakly held to the clay mineral surfaces, and can be readily displaced by other cations present in the solution. The cation (base) exchange depends on the ionic strength (charge to ionic radius ratio) and the concentration of the solution. The cation exchange (or adsorption) capacity of a clay mineral reflects its charge deficiency per unit mass and is a function of the particle composition and of its specific surface (Table 7.4).

Clay particles are held in suspension by electrostatic forces that act between their surface charge and that of the ions of the solution, thus resisting settling by gravity. The interaction between any pair of clay particles is governed by the net result of the balance between their attractive forces primarily van der Waals curve 4 (inversely proportional to the square of the separation distance), and the repulsive electrostatic forces curve 1 (decreasing exponentially with increasing distance of separation) (Figure 7.3). If the energy provided to the system (by the salt ions) is less than the maximum of the repulsive curve 3 in Figure 7.3 (the barrier energy), the particles will

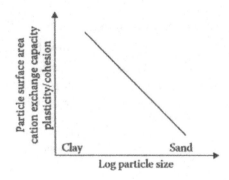

FIGURE 7.2 Particle surface area and other properties versus particle size.

(Chaney and Almagor, 2016. Republished with permission of Taylor & Francis)

separate, or disperse, spontaneously with loss of the potential energy of the system. If, however, the energy required to bring the particles closer than the distance corresponding to the energy barrier is available, the particles will move rapidly toward each other, or flocculate. This movement will result in a release of energy, as the particles move to a separation distance dictated by the born repulsive forces between the clay particles, which correspond to the minimum of curve 2. Born repulsive force develops at contact points between particles, resulting from the overlap between electron clouds. It is sufficiently great to prevent the interpenetration of matter. At separation distances beyond the region of direct physical interference between adsorbed ions and between hydration water molecules, double-layer interactions provide the major source of interparticle repulsion. Particle separation is more difficult to achieve than

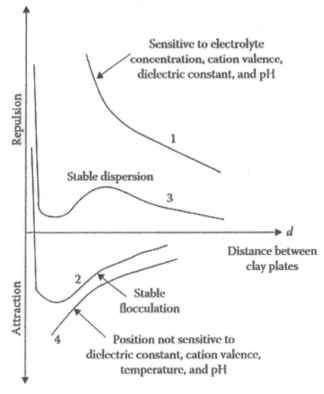

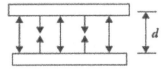

FIGURE 7.3   Relationship between the forces of particle interaction, F_i and distance between particles.

(Chaney and Almagor, 2016. Republished with permission of Taylor & Francis)

particle aggregation, because the amount of energy required to separate the particles is much larger than the energy needed to bring them together. Dispersion in seawater occurs where water turbulence prevents contact between clay particles or in the presence of peptizers, such as dissolved phosphates carried by rivers. The tendency toward flocculation increases as the suspended particle concentration, the electrolyte concentration in the solution, the valence of ions, the temperature increase, and the dielectric constant (size of hydrated ions, pH, and anion adsorption) decrease. Likewise, water turbulence can increase interparticle collisions, thus providing the force required to overcome the potential energy barrier that prevents flocculation (Kranck, 1980). Flocculation also is enhanced in the presence of organic matter (>5%). Readily attached to silicate minerals and adsorbed by water molecules, organic matter effectively reduces the net surface charge of the clay particles. Thus, the presence of organic matter allows the clay particles to approach one another more closely, and in addition promotes particle-binding and bonding by polymeric products of microbial metabolism and within interwoven plant fibers (Bennett and Hulbert, 1986; O'Brien, 1970; Pusch, 1973a, 1973b; van der Ven, 1981). As more and more particles flocculate, the agglomerates become large and settle by gravity, forming a sediment on the seafloor in which the interparticle pores are filled with seawater.

7.5.4 CLAY MICROSTRUCTURE

The chemical nature of the fluid medium of saline water strongly affects the electrochemical interactions between suspended clay particles, thus determining the fabric characteristics of dispersion or flocculation prior to deposition. In non-saltwater the clay particles remain dispersed, settling slowly in quiet water and forming open networks of more or less uniformly distributed, largely parallel-oriented particles with moderately high void ratios. However, on contact with saline seawater the clay particles in suspension rapidly flocculate, producing voluminous agglomerates, several micrometers in width, of randomly oriented particles (flocs or flocules) (Figure 7.4). These flocs in turn tend to become denser and smaller as the water salinity decreases (Keller, 1957; Mitchell, 1956; van Olphen, 1977; Whitehouse et al., 1960).

(a)   (b)   (c)

FIGURE 7.4 Various fine-grained sediment structures: (a) undisturbed saltwater sediment (flocculated); (b) undisturbed freshwater sediments (partially flocculated); and (c) remolded sediment (dispersed).

(Chaney and Almagor, 2016. Republished with permission of Taylor & Francis)

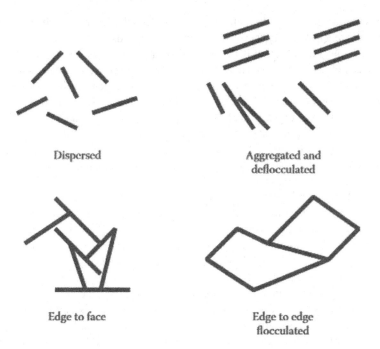

FIGURE 7.5   Modes of particle associations in clay suspensions and terminology.

(Chaney and Almagor, 2016. Republished with permission of Taylor & Francis)

Particle associations in clay suspensions can be described as follows and as illustrated in Figure 7.5 (van Olphen, 1977): (1) dispersed: no face-to-face (FF) association of clay particles; and (2) aggregated: FF association of several clay particles.

Larger and thicker particles are the result of FF association. EF and EE associations, by contrast, can produce a card house type of structure that is quite voluminous until compressed.

The terms flocculated and aggregated are often used synonymously in a generic sense to refer to multiparticle assemblages. Similarly deflocculated and dispersed are used synonymously in a generic sense to refer to single particles or small particle groupings acting independently.

Individual clay sheets normally stack in parallel, FF, or stepped-FF alignment, forming packets or domains (Aylmore and Quirk, 1959; Bennett et al., 1977; Moon, 1972). Single particles are less common. Bonding between the sheets, typically of a covalent character, is modest or weak. Different clays experience different types of sheet bonding: electrostatic attraction to intersheet potassium (muscovite, illite) and magnesium ions (vermiculite), hydrogen bonding between oxygen ions and hydroxyl groups (kaolinite), van der Waals forces by water molecules (Halloysite) or exchangeable cations (montmorillonite), and so on. The strength of the bonding between the clay sheets, however, differs markedly from one clay mineral to another, resulting in a highly variable response to changes in the physiochemical environment. Among the major clay groups, the kaolinite-related minerals are little affected in neutral aqueous media because of the tight hydrogen bonding between their structural units. This results in a restricted surface area; consequently, adsorptive capacity for cations and water molecule attraction is limited, as revealed by a very low plasticity (capacity to be remolded), cohesion, shrinkage, and swelling properties. Kaolinite crystals are relatively large, often comprising nearly perfect hexagons with clean-cut edges. The weakly oxygen-to-oxygen linked, tiny crystal sheets of smectites easily break down into individual crystal sheets. This occurs when polar water molecules enter between the adjacent sheets and separate them to such an extent that van der Waals force is no longer sufficient to hold them together. The clay sheet separation in smectites results in a very large surface area per unit mass of soil, and subsequent surface attraction of swarms of cations.

The high cation adsorption capacity of smectites, 10–15 times that of kaolinite, is responsible for their high plasticity, cohesion, and marked shrinkage on drying. By contrast, because the strong potassium ion bonding of their crystal units prevents extensive hydration and cation adsorption, illite clays are relatively nonexpansive, possessing intermediate plasticity, swelling, and shrinkage properties between kaolinite and smectites.

Attraction between positively charged edges of the face to face (FF) -held domains and negatively charged surfaces of others produces stable edge to face (EF) links, which maintain the relative rigidity of the complex structure of the particle aggregates (Figure 7.5). Edge to edge (EE) links between stepped face to face (FF) domains and individual particles form chains of varying lengths and thickness.

Clay microfabric refers to the spatial distribution, orientation, and particle-to-particle relationship of the solid particles of sediments (Bennett and Hulbert, 1986). The particle network of submarine fine-grained sediments consists of particle aggregates and minor quantities of quartz, feldspar, and mica; opaque heavy minerals; skeletal carbonate fragments; and plant debris, normally arranged in highly porous, loose, and random patterns and linked by chains. In smectite-rich sediments, amorphous smectite material, appearing as continuous fleecy-looking fabric, tends to cluster around and between the clay domains and individual particles. Up to 30% to 50% of the total mass of most of the fine-grained deposits consists of either silt- and clay-sized, highly indurated shale clasts, soft lumps, and crumbs (large aggregates showing finite shape and structure integrity) of various shapes, sizes, and origin (fluvial, windblown, and slump-released) that are randomly distributed within the clay matrix (Bryant and Bennett, 1988). The chains and flocs provide the necessary strength (cohesion) of the sediments that resists deformation under external stress. Bryant and Bennett (1988) argue that the large clasts within the sedimentary mass, which are supported by the chains and flocs (in TEM photomicrographs they appear to lack support because of ultrathin sectioning of samples) act as filters which shorten the links thus strengthening the fabric. Substantial porosity is created by the edge to face (EF) particle contacts and the edge to edge (EE) -linked chains. The degree of openness of the fabric is largely governed by the mineralogy and the size of the clay particles, and by the amount and shape (angularity) of silt particles in the sediments. The finer the particles, the smaller the voids, and the greater their number. This results in voids that are mostly discontinuous or narrowly interconnected, which results in high porosities and relatively low permeabilities in clayey sediment(s). Although thin-shelled and extremely open, the (micro)fabric of clay masses is sufficiently rigid to impart origin strength to the highly porous, water-saturated soils formed at the sediment–water interface (Jakobson, 1953), even in areas of high sedimentation rates. The rigidity of the fabric would explain the apparent over-consolidation of numerous superficial, highly porous submarine sediments (Mathewson in a personal communication to Bennett and Hulbert, 1986), especially on the deep ocean floor where the sedimentation rates are very small (~1 mm/1000 years). Resistance to shearing (erosion) by moving water is yet another manifestation of the strength of the clayey fabric. On the other hand, remolding causes breakdown of aggregates, displacement of particles, and destruction of the bonds between them. This results in an improvement in parallel orientation of the clay particles (homogenization of the soil), decrease in void ratio, and loss of the soil's shear strength.

Once formed, the electrochemical interactions play a very passive role in the post depositional changes in the clay fabric. The clay fabric after deposition becomes dominantly controlled by mechanical processes (action of waves, currents and tides, sediment overloading, seismic shocks, and biological activity) (Bennett, 1976; Bennett and Hulbert, 1986). Under (relatively) low stresses during the first stages of consolidation, the aggregates appear to initially move as units in connection with slight deformation of their links (chains and domain contacts). This results in denser packing; yet the overall random appearance of the microfabric and its relatively high void ratio are preserved. As the stress increases, the particle adjustment results in denser particle-to-particle packing through an increase in the degree of particle orientation and reduction of the total volume of the voids. Under high stress, distinct distortions of the original microfabric develop: particle contacts break, crumbs and aggregates distort and/or disintegrate, and the particles are brought into close contact. Thus, the following occurs: particle reorientation is accentuated and a great degree of parallelism (preferred orientation) is achieved; long chains and stepped face-face (FF) -arranged domains form layers; the voids become long, thin, and narrow, trending in parallel to the particle orientation; and long dewatering channels develop. In most cases, however, these changes are not related directly to the applied stress (depth of burial) and vary considerably from sediment to sediment. Furthermore, the degree of preferred orientation varies considerably within different sections of one soil that was subjected to a uniform external stress.

The effects of post-depositional processes on the microfabric of the sediments is poorly known. These effects include biological activity (bioturbation), gas production, intrastratal geochemical changes (salinity gradients), and authigenic mineralization, such as carbonate cementation and bacterially deposited agglomerates of iron sulfides in the pore spaces.

Alteration of the microfabric by these processes can result in significant changes in the porosity and the geotechnical properties of the sediments (Shorten, 1993). Even during the early stages, carbonate cementation (carbonate bonding) probably strongly restricts particle rearrangement during consolidation, resulting in a large degree of apparent consolidation (Nacci et al., 1974; Shorten, 1995).

7.5.5 CONSISTENCY PROPERTIES

The plasticity characteristics of a sediment is characterized by its consistency. Atterberg limits and related indexes are a function of many parameters: drying (Casagrands, 1932), temperature, molding and grain size (White and Walton, 1937), surface area (Farrar and Coleman, 1967), and clay mineral composition (Seed et al., 1964). The main factors are those tied to the physicochemical properties of the clay–water system (Moum and Rosenquvist, 1961; Soderblom, 1969).

Generalized plasticity characteristics of sediments from a variety of locations (prodeltas, deltas, gulfs and bays, continental margins and trenches, etc.) are presented in Figure 7.6 (Chassefiere and Monaco, 1983). A review indicates that both the liquid limit (I_L) and the plasticity index (P_I) increase with increasing smectite content. This behavior is centered around the A-line on the Casagrande liquid limit chart. An increase in the percentage of organic matter results in the decrease of P_I to the approximate upper limit exhibited by peats.

Organic matter is a common constituent of marine sediments; it frequently constitutes up to 2% or more by the weight of sediments, especially on continental margins. It exists as biota, including living organisms such as microbes and macroscopic infauna, and as detritus derived from an innumerable variety of dead organisms. It exists in a variety of gross forms including filaments, mats, and particles, and in a variety of chemical forms, including proteins, carbohydrates, lipids, and lignins. Organic matter is thus a general term and is used herein to describe all organic and organic-related components. A summary of much of the information on organic matter has been presented by Hunt (1979).

The effect of different types and amounts of organic matter on soil properties has been the subject of a number of investigations. A thorough study of some sensitive Swedish clays and other Swedish soils by Pusch (1973a) has detailed the complexity of the possible organic matter–clay relationships and demonstrated the profound impact that organic matter can have on the index properties, deformational characteristics, and strength parameters of a soil. Pusch noted from his analytical work that progressive failure and deformation were observable at an organic content of 2% to 3% and that shear strength was significantly affected at an organic content of 3% to 4%. In addition, Pusch showed a dramatic increase in both the liquid and the plastic limit with increasing organic content. This phenomenon was also observed by Odell et al. (1960) using samples from several soil groups and from different soil horizons in Illinois.

Similar effects and relationships have been noted for marine sediments. In a study of organic-rich sediments (up to 20% organic carbon) on the Peru–Chile continental slope, Busch and Keller (1981) showed an increase in both liquid and plastic limits with increasing organic content. Reimers (1982) analyzed these same sediments and the different types of organic matter that were present. She suggested that changes from one type of organic matter to another during humification could be reflected in the geotechnical properties, specifically that mobilization of organic matter may result in a decrease in compressibility.

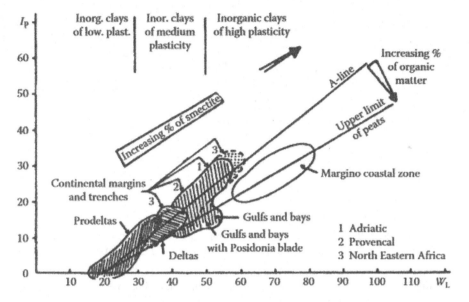

FIGURE 7.6 Activity diagram of freshwater and saline water. (From Chassefiere and Monaco, 1983. Reproduced with permission of Taylor & Francis Group)

7.6   COARSE-GRAINED SOILS

7.6.1   INTRODUCTION

The majority of beach sand are grains of quartz, a simpler and chemically more inert material than clay minerals. In the geologist's size classification, sand grains range from 16 to 500 times larger in diameter than the largest clay particle (4000 to more than 100 million times larger in volume). At this size, the force of gravity acting on individual sand grains exceeds the surface forces. Therefore, the surface forces acting on sand are far less important than on clay particles. Because sand grains are not controlled by surface forces they do not exhibit cohesion (i.e., they do not stick together).

7.6.2   CLASSIFICATION OF SEDIMENT BY SIZE

7.6.2.1   Particle Diameter

The size of the sediment particles is one of its most important characteristics (Figure 7.7). The range of grain sizes covering about seven orders of magnitude is of practical interest to coastal engineers. This range of particle size ranges from clay to large breakwater armor stone blocks. The size of a particle size is typically defined in terms of its diameter. Use of the term diameter can be ambiguous since grains are typically irregularly shaped. The diameter of a cohesionless particle >0.002 mm is usually determined by the mesh size (i.e., opening size) of a sieve that will allow the grain to pass. This

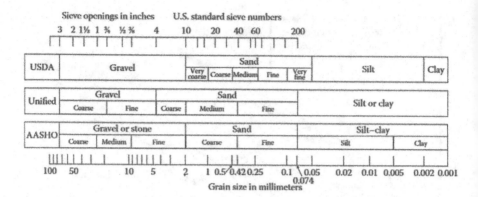

FIGURE 7.7 Comparison of particle size scales.

(Data from Bureau of Reclamation, *Earth Manual, Part 1*, 3rd ed., U.S. Department of the Interior, U.S. Government Printing Office, Washington, DC, 329, 1998)

opening size of the sieve is defined as a particle's diameter. When performed in a standard manner such as using ASTM D2487 provides repeatable results.

For finer grain material it's diameter is normally determined by using a hydrometer analysis (ASTM D422) to determine its diameter by its fall velocity. A grain's sedimentation diameter is the diameter of a sphere having the same density and fall velocity. This definition has the advantage of relating a grain's diameter to its fluid behavior.

There is a need to determine a diameter for an aggregation of particles, rather than the diameter of a single particle. Natural sediments generally have a range of grain sizes. Sediment samples found in nature typically have a small number of large particles and a large number of smaller-diameter particles. This collection

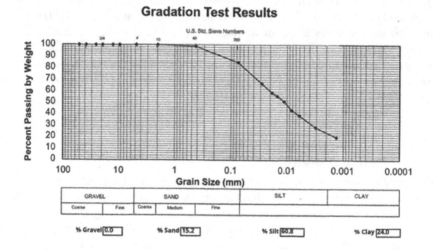

FIGURE 7.8 Example of sediment distribution using semilog paper, Humboldt Bay, CA, Entrance Channel.

of particle size fractions can be described using a logarithmic scale (Figure 7.8). If each size fraction is weighed, then a typical sediment sample will have a normal distribution.

Plotting the sediment sample fractions as a percentage of the total weight of the material being sieved versus the size distribution results in an approximate straight line on a log-normal graph. Standard statistical parameters can be used to describe the curve.

7.6.2.2 Sediment Size Classifications

Sediment sizes have arbitrarily been broken down into classes such as cobbles, sand, silt. A number of methods have been proposed. Two classification systems are currently in general use. These two methods are the Modified Wentworth Classification, and Unified Soils Classification or the ASTM Classification systems.

7.6.2.2.1 Modified Wentworth Classification

The Modified Wentworth Classification system is generally used in geologic work. This is because geologists are typically interested in particle size as indicating processes or events (Blatt et al., 1980).

7.6.2.2.2 Unified Soils Classification or the ASTM Classification

The Unified Soil Classification system was developed by various American engineering groups. These groups have been concerned with standardizing test procedures to obtain repeatable results in the analyses of sediment (Volume 4.08 of the Standards published by the American Society for Testing and Materials (ASTM)).

These two systems are compared in Table 7.5. Note that there are differences in the category limits. It is necessary to indicate which classification system is being utilized when describing a sediment.

7.6.2.3 Units of Sediment Size

7.6.2.3.1 Coarse Grain Material

There are three ways to describe the size of a coarse grain sediment particle: U.S. Standard sieve numbers, millimeters, and phi units. A sieve number is approximately the number of square openings per inch, measured along a wire in the wire screen cloth (Tyler, 1991). The millimeter measurement dimension is the length of the inside of the square opening in the screen cloth. This measurement is the nominal approximation to the actual sediment size. The Wentworth scale has divisions that are whole powers of 2 mm as shown in Table 7.5. The property of using 2 mm as class limits led Krumbein (1936) to propose a phi unit scale based on the definition:

$$\varphi = \log_2 D, \tag{7.4}$$

where D is particle diameter in millimeters.

Phi diameters are indicated by writing φ after the numerical value. As an example, a 2.0-φ sand grain has a diameter of 0.25 mm. To convert from phi units to millimeters, the inverse equation is used:

$$D = 2^{-\varphi}. \tag{7.5}$$

TABLE 7.5
Sediment Particle Sizes

ASTM (Unified) Classification[1]	U.S. Std. Sieve[2]	Size in mm	Phi Size	Wentworth Classification[3]
Boulder		4096	−12.0	Boulder
		1024	−10.0	
Coarse gravel	12 in. (300 mm)	256	−8.0	Large cobble
		128	−7.0	Small cobble
		107.64	−6.75	
		90.51	−6.5	
Fine gravel	3 in. (75mm)	76.11	−6.25	
		64.00	−6.0	Very large pebble
		53.82	−5.75	
		45.26	−5.5	
		38.05	−5.25	
		32.00	−5.0	Large pebble
		26.91	−4.75	
		22.63	−4.5	
Fine gravel	¾ in. (19 mm)	19.03	−4.25	
		16.00	−4.0	Medium pebble
		13.45	−3.75	
		11.31	−3.5	
		9.51	−3.25	
		8.00	−3.0	Small pebble
		6.73	−2.75	
		5.66	−2.5	
Coarse sand	4 (4.75 mm)	4.76	−2.25	
		4.00	−2.0	Granule
		3.36	−1.75	
		2.83	−1.5	
		2.38	−1.25	
Medium sand	10 (2.0 mm)	2.00	−1.0	Very coarse sand
		1.68	−0.75	
		1.41	−0.5	
		1.19	−0.25	
		1.00	0.0	Coarse sand
		0.84	0.25	
		0.71	0.5	
		0.59	0.75	
		0.50	1.0	Medium sand
Fine sand	40 (0.425 mm)	0.420	1.25	
		0.354	1.50	
		0.297	1.75	

(Continued)

TABLE 7.5 *(Continued)*
Sediment Particle Sizes

ASTM (Unified) Classification[1]	U.S. Std. Sieve[2]	Size in mm	Phi Size	Wentworth Classification[3]
		0.250	2.0	Fine sand
		0.210	2.25	
		0.177	2.5	
		0.149	2.75	
		0.125	3.0	Very fine sand
		0.105	3.25	
		0.088	3.5	
Fine-grained soil	200 (0.075 mm)	0.074	3.75	
		0.0625	4.0	Coarse silt
		0.0526	4.25	
		0.0442	4.5	
		0.0372	4.75	
		0.0312	5.0	Medium silt
		0.0156	6.0	Fine silt
		0.0078	7.0	Very fine silt
		0.0039	8.0	Coarse clay
		0.00195	9.0	Medium clay
		0.00098	10.0	Fine clay
		0.00049	11.0	Colloids
		0.00024	12.0	
		0.00012	13.0	
		0.000061	14.0	

Note: (1) ASTM Standard 2487-92. This is the ASTM version of the Unified Soil Clssification System. Both systems are similar from ASTM (1994). (2) Note that British Standard, French, and German DIN mesh sizes and classifications are different.

Source: U.S. Army Corps of Engineers (2008). Courtesy of the U.S. Army.

The advantages of utilizing the phi unit in the Wentworth scale system include the following:

a. Whole numbers occur at the limits of sediment classes.
b. System allows comparison of different size distributions because it is dimensionless.

In contrast, the disadvantages of the phi unit in the Wentworth scale system are the following:

a. Phi units get larger as the sediment size gets smaller.
b. Difficult to physically interpret particle size in phi units.
c. Since it is a dimensionless unit, the phi unit cannot represent a length.

7.6.2.4   Median and Mean Grain Sizes

All natural sediment samples contain particles with a range of sizes. At times it is useful to describe the sediment sample using a single parameter. A number of parameters have been utilized. The median particle diameter (Md) of the sample is a parameter that is often utilized. The definition of Md is the weight of one-half of the particles in the sample by size.

The median diameter can also be written as D_{50}. Other size fractions can also be similarly determined. For example, D_{90} is the particle diameter for which 90% of the sediment, by weight, has a smaller diameter. An equivalent definition holds for the median of the phi-size distribution $\varphi 50$ or for any other size fraction in the phi scale.

The mean grain size is another measure of a sediment sample. To compute this quantity, several formulas have been proposed. These formulas typically use a cumulative size distribution plot of the sample (Folk and Ward, 1957; Inman, 1952; McCammon, 1962; Otto, 1939). These formulas typically use averages of 2, 3, 5, or more selected percentiles of the phi frequency distribution. Following Folk (1974):

$$M_\phi = \frac{\left(\phi_{16} + \phi_{50} + \phi_{84}\right)}{3}, \tag{7.6}$$

where M_ϕ is the estimated mean grain size of the sample in phi units.

The median and mean grain sizes are normally similar for the majority of beach sediments (Ramsey and Galvin, 1977). If the particle sizes in a sample are log-normally distributed, the two measures are identical. Since the median is easy to determine it is normally used to characterize the central tendency of a sediment sample.

7.6.2.5   Higher-Order Moments

Variation of the sediment sample distribution from a log-normal distribution can be determined using additional statistics. These additional statistical measures are (1) standard deviation, (2) skewness, and (3) kurtosis.

The standard deviation is a measure of the how much the sample spreads out around the mean (i.e., its sorting). The standard deviation can be approximated following Folk (1974) by:

$$\sigma_\phi = \frac{\left(\phi_{84} - \phi_{16}\right)}{4} + \frac{\left(\phi_{95} - \phi_{5}\right)}{6}, \tag{7.7}$$

where σ_ϕ is the estimated standard deviation of the sample in phi units.

For a completely uniform sediment ϕ_{05}, ϕ_{16}, ϕ_{84}, and ϕ_{95} are all the same. The standard deviation for this case is zero.

There are also additional descriptions of the standard deviation. A sediment is described as well sorted if all particles have sizes that are close to the typical size (i.e., small standard deviation). If the particle sizes of the sample are distributed evenly over a wide range of sizes, then it is well graded. A well-graded sample is therefore poorly sorted. In contrast, a well-sorted sample is poorly graded.

The symmetry of the distribution is measured by the phi coefficient of skewness α_ϕ as defined in Folk (1974) as:

$$\sigma_\phi = \frac{\phi_{16} + \phi_{84} - 2(\phi_{50})}{2(\phi_{84} - \phi_{16})} + \frac{\phi_5 + \phi_{95} - 2(\phi_{50})}{2(\phi_{95} - \phi_5)}. \tag{7.8}$$

The skewness is zero for a perfectly symmetric distribution. In contrast, a positive skewness indicates there is a tailing out toward the fine sediments, and conversely, a negative value indicates more particles are coarser.

The sharpness of the distribution coefficient is the kurtosis β_ϕ. This parameter is a measure of the proportion of the sediment in the middle of the distribution relative to the amount in both tails. Following Folk (1974), kurtosis is defined as:

$$\beta_\phi = \frac{\phi_{95} - \phi_5}{2.44(\phi_{75} - \phi_{25})}. \tag{7.9}$$

Mean and median particle size values are frequently converted from phi units to linear measures. In contrast, the standard deviation, skewness, and kurtosis should remain in phi units because they have no corresponding dimensional equivalents. If these terms are used in equations, they should be used in their dimensionless phi form. Relative relationships are given for ranges of standard deviation (Table 7.6), skewness (Table 7.7), and kurtosis (Table 7.8).

TABLE 7.6

Qualitative Sediment Distribution Ranges for Standard Deviation

Phi Range	Standard Deviation Description
<0.35	Very well sorted
0.35–0.50	Well sorted
0.50–0.71	Moderately well sorted
0.71–1.00	Moderately sorted
1.00–2.00	Poorly sorted
2.00–4.00	Very poorly sorted
>4.00	Extremely poorly sorted

TABLE 7.7

Qualitative Sediment Distribution Ranges for Skewness

	Coefficient of Skewness
<0.3	Very coarse-skewed
−0.3 to −0.1	Coarse skewed
−0.1 to +0.1	Near symmetrical
+0.1 to +0.3	Fine skewed
>+0.3	Very fine skewed

TABLE 7.8

Qualitative Sediment Distribution Ranges for Kurtosis

	Coefficient of Kurtosis
<0.65	Very platykurtic (flat)
0.65–0.90	Platykurtic
0.90–1.11	Mesokurtic (normal peakness)
1.11–1.50	Leptokurtic (peaked)
1.50–3.00	Very leptokurtic
>3.00	Extremely leptokurtic

7.6.2.6 Uses of Particle Size Distributions

The most commonly used sediment size characteristic is the median grain size. The second parameter is the standard deviation of sediment sample distribution. This parameter has been used in a number of ways. These uses include beach-fill design (see Hobson [1977], and sediment permeability, Krumbein and Monk [1942]). There are a number of papers in the literature on potential applications utilizing various statistical measures on the size distribution; Folk (1965, 1966), Folk and Ward (1957), Griffiths (1967), Inman (1957), McCammon (1962), and Stauble and Hoel (1986).

7.6.2.7 Sediment Sampling Strategies

A beach changes in time and space due to the composition of the available sediments. As an example, beach sediment distributions in winter are typically coarser and more poorly sorted than in summer. Also, typically, there is a larger variability in both the foreshore and the bar/trough regions than in the dunes and the nearshore.

To characterize the sediments a set of samples is usually obtained at a site. To reduce the high variability in special grain size distributions on beaches samples are combined from across the beach (Hobson, 1977). Composite samples are created by either: (1) physically combining several samples before sieving or (2) by mathematically combining the individual sample weights. The set of samples obtained can be small if the purpose is only to characterize the beach as a whole. However more samples are needed, if the purpose is to compare and contrast different portions of the same beach. In this case a sampling scheme prior to fieldwork is usually necessary to develop.

Samples should be collected at all major changes in morphology along the profile in a cross-shore sampling program. The areas to be sampled are as follows: (1) dune base, (2) mid-berm, (3) mean high water, (4) mid-tide, (5) mean low water, (6) trough, (7) bar crest (Stauble and Hoel, 1986). Sediment sampling should coincide with survey profile lines in the longshore direction. The samples should be spatially located and related to morphology and hydrodynamic zones. Engineering structures should be considered in choosing sampling locations along with shoreline variability.

Typically a sampling line should be spaced every half mile. Engineering judgment is required to define adequate project coverage.

Samples collected along similar depositional energy levels can be combined into composite groups of profile sub-environments. The most usable composites to characterize the beach and nearshore environment are intertidal and subaerial beach samples. A composite sample containing material from the mean high water, mid-tide, and mean low water have been found to give the best representation of the foreshore beach (Stauble and Hoel, 1986).

7.6.2.8   Laboratory Procedures

Several techniques are available to analyze the size of beach materials. Each technique is restricted to a range of sediment sizes. Material in the pebbles and coarser material sizes are usually measured with calipers. For smaller sediments you can use coarse sieves for material up to about 75 mm.

Sand-sized particles are usually analyzed using sieves (i.e., medium gravels through coarse silt). This requires a stack of sieves of square-mesh woven-wire cloth that are ordered from smallest opening at the bottom to the largest at top. Each sieve in the stack differs from the adjacent sieves by having a nominal opening less than the opening of the sieve above it and greater than the opening of the sieve below it. A pan is placed below the bottom sieve. The sample is poured into the top sieve, a lid is placed on top, and the stack is placed on a shaker, usually for about 15 minutes. The various grains fall through the stack of sieves until each particle reaches a sieve that is too fine for it to pass. Then the amount of sediment in each sieve is weighed. ASTM Standard D422 is the basic standard for particle size analysis of soils. Sample preparation for D422 is described in Standard D421.

Sieves are graduated in size of opening according to the U.S. Standard series or according to phi sizes.

Sieve openings range must span the range of sediment sizes to be sieved. Typically, analysis of a particular sediment requires approximately 6 full-height sieves or 13 half-height sieves plus a bottom pan.

If 6 sieves are used, each typically varies in size from its adjacent neighbors by a half phi; if 13 sieves are used, they usually vary by a quarter phi. Normally, approximately 40 g of sediment is sieved. For large size fractions typically more material is needed (ASTM Standard D2487).

Sediment size fractions for silts and clays are usually not necessary. Usual practice is normally to note that a certain percentage of the material are fines with diameters smaller than the smallest sieve. When measurements of fines are required, either the pipette or the hygrometer method (ASTM D422) is usually used. Both of these methods are based upon determining the amount of time that different size fractions remain in suspension. The pipette method is discussed in Vanoni (1975). Coulter counters have also been used occasionally.

7.6.2.9   Particle (i.e., Grain) Shape and Abrasion

The particle shape is important to coastal engineers because it affects several other properties. When the grains are far from spherical it effects the fall velocity, sieve analysis, initiation of motion, and certain bulk properties, such as porosity and angle of repose.

Particle (i.e., Grain) shape is primarily a function of its composition, size, original shape, and weathering history. The shape of littoral material ranges from nearly spherical (e.g., quartz grains) to nearly disklike (e.g., shell fragments, mica flakes) to concave arcs (e.g., shells). Classifying sediment particle shape has been divided into three parts: (1) the sphericity or overall shape of a particle, (2) the roundness or the amount of abrasion of the corners, and (3) the particle roughness.

However, most littoral grain shapes are close to spheres that a detailed study of their shape is not warranted. Therefore, a qualitative description of roundness is sufficient. This can be accomplished by comparing the grains in a sample to photographs of standardized grains (see Krumbein, 1941; Powers, 1953; Shepard and Young, 1961).

Studies of sand grain abrasion have been conducted because of the worry that abrasion of beach sand contributes to beach erosion. These investigations found that abrasion of the typical quartz beach sand is rarely significant. In general, recent studies support the conclusion of Mason (1942) that on sandy beaches the loss of material due to abrasion occurs at rates so low as to be of no importance in shore protection problems.

For quartz to abrade very high stresses due to impact forces are needed. These forces are developed by sudden changes in momentum. Momentum is the product of mass and velocity. The small mass of a sand particle generates large forces only by grains moving at high velocities. The drag on a sand particle moving in water increases as the square of its velocity, which limits its velocity to a low multiple of its fall velocity. Because fall velocities are only a few centimeters per second, therefore, it is difficult to achieve a stress between impacting grains that is anywhere near the strength of quartz. Thus, the rounding of the corners of angular quartz grains in river or littoral environments is a very lengthy process. Sands, silts, and clays can typically be considered as the end product of the weathering process of rocks.

Large particles such as boulders and riprap subject to wave action commonly experience abrasion. This is because the mass of a particle increases with the cube of its diameter. If that boulder was a perfect sphere, and rested with a point contact on the plane surface of another rock. Then merely the weight of the boulder would crush the point contact of the sphere. This crushing would continue until the area of contact increased enough to reduce the pressure of the contact to below the crushing strength of the boulder. Thus, material is abraded from the boulder crushed at this contact point. This process of stress concentration at points of contact has been considered quantitatively by Galvin and Alexander (1981).

If the boulder moves with motion imparted by the arrival of a wave crest, the slight velocity of the boulder mass produces a momentum. This momentum can produce impact forces in excess of crushing strength at points of contact between a boulder and its neighbors, thus abrading the rock.

A large rock typically will break along surfaces of weakness, the pieces after breakage will usually be stronger than the material from which the pieces are broken. Thus, abraded gravel pieces on a wave-washed shingle beach usually have greater strength than the bulk strength of the rock from which the gravel was derived.

One interest of particle shape to coastal engineers is in the design of man-made interlocking armor units on breakwaters that have high stability, even when stacked

TABLE 7.9
Typical Maximum and Minimum Void Ratios, Porosities, and Unit Weights for Various Soils

Description	Void Ratio		Porosity (%)		Dry Unit Weight (kN/m)³	
	emax	emin	nmax	nmin	$\gamma_d$min	$\gamma_d$max
Uniform spheres	0.92	0.35	47.6	26.0	–	–
Standard Ottawa sand	0.80	0.50	44	33	14.5	17.3
Clean uniform sand	1.0	0.40	50	29	14.0	170
Uniform inorganic silt	1.1	0.40	52	29	12.6	18.5
Silty sand	0.90	0.30	47	23	13.7	20.0
Fine to coarse sand	0.95	0.20	49	17	13.4	21.7
Micaceous sand	1.2	0.4	55	29	11.9	18.9
Silty sand and gravel	0.85	0.14	46	12	14.0	22.9
Soft clay	1.4	0.9	58	47	11.5	14.5
Soft organic clay	3.2	2.5	76	71	6	8
Stiff clay	0.6	–	38	–	17	–

at a high angle of repose. Particle shape has also been used to estimate residence time in the littoral environment. See Krinsley and Doornkamp (1973) and Margolis (1969) (Table 7.9).

Particles having low sphericity, such as flat gravels and flaky micas, often construct arches within the soil that bridge over large voids, resulting in loose, yet tightly wedged, stable arrangements (Figure 7.9). These particles also may tightly interlock and form dense packing configurations. Flaky particles often form oriented structures, developed during sedimentation or resulting from movement by shear stresses or high pressure.

Both loose and dense soil structures are capable of supporting considerable static overburden loads with little or no distortion. However, loose soils with grain sizes in the coarse silt fine sand range are inherently metastable as shown by Terzaghi (1925). Shocks and vibrations readily cause movement of the particles to denser, more stable arrangements. Highly metastable, loose sediments are often produced by rapid deposition of silt and fine sand in quiet environments, such as delta fronts of large rivers. Many of the fine silt-sized particles are actually land-derived minuscule mudstone (indurated clays) particles, while others of nonclayey compositions have an adhering film of clay. Therefore, silts can possess some clayey properties, such as plasticity, cohesion, and adsorption. The structure of these sediments is sustained by cohesive forces at the silt grain contact and the friction at the contact of the sand grains directly perched on one another. This structure readily collapses upon shaking, which causes complete loss of their grain-to-grain contacts (liquefaction), followed by settlement of the suspended grains, displacement of their pore water, and packing to a higher density.

The porosity does not indicate the degree to which a soil is loose or dense. Porosity is defined as the ratio of the volume of the voids in a given mass of soil to its total

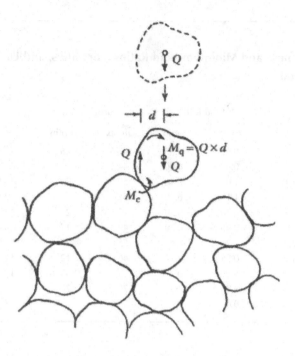

FIGURE 7.9 Loose, metastable structure showing adhesion at point of first contact between sinking particle and surface of sediment prevents it from rolling into stable position.

(Data from Terzaghi [1956])

volume. The degree of compaction, termed relative density Dr, of a cohesionless soil can be quantitatively obtained by comparing the porosity of the soil with its loosest and densest possible states. This concept has been previously discussed in Section 7.4.2 and will not be repeated here.

7.7   CLASSIFICATION OF SEDIMENTS

7.7.1   BASES OF SEDIMENT CLASSIFICATION

A number of sediment properties are important in coastal engineering. Most of these properties can be placed into one of three groups: (1) the size of the particles making up the sediment, (2) the composition of the sediment, and (3) bulk characteristics of the sediment mass.

There are strong correlations among the three classification groups. Clay particles are small and platey as discussed previously. Because the surface area of a clay particle is so large it is chemically active. The aggregate of clay surfaces when wet produces the cohesive, plastic, and slippery characteristics of its bulk form.

In contrast, most grains of beach sand are massive quartz particles, and chemically inert when compared to clay minerals. Therefore, gravity forces control the behavior of sand grains. Several differences between clay and sand are summarized in Table 7.3. More inclusive discussions of sediment sizes, compositions, and bulk properties are given later in this chapter.

Sediment properties of material existing at a site, or that might be imported to the site, have important implications for any coastal engineering project. The following sections briefly discuss several systems used to classify soils with similar properties.

7.7.2 Systems of Classification

Soil classification deals with the categorization of soils based on distinguishing characteristics as well as potential criteria that dictates choices in use. Soil classifications arise from the need for (1) rational, yet convenient characterization of soils; (2) their correlation (differentiation, comparison, and contrast); and (3) methods for effective communication based on a uniformity of criteria and nomenclature.

It is difficult to classify soils because their gradations differ from one form to another. This difference is caused by a combination of the frequent mixing of sediments of different size, composition and origin, and changes caused by a variety of transportational and diagenetic processes. Therefore, the existing classifications are constructed on the basis of a few significant properties of the sediments. Several classification systems of sediments have been proposed based on their chemical and mechanical composition, selected geotechnical properties, agencies of transport and deposition, and sedimentary environments.

Difficulties, however, still remain, mostly because the significant properties of one sediment group according to any classification system are not necessarily the significant properties of another. Even very partial classifications based on a single, well-defined significant property (e.g., permeability, compressibility) evaluated to address specific types of problems have serious limitations. As an example, the most commonly used classification system based on grain size does not consider the chemical and mineralogical composition of the soil particles even though these properties govern particle hardness and density, reaction to pore fluids, and so on. Consequently, unified classification systems evaluated to simultaneously employ a wide range of significant properties must either employ descriptive terms (e.g., the general classification of marine sediments—Figure 7.10), or be accompanied by auxiliary descriptive charts (as is done in the Unified Soil Classification System [USCS], Wagner [1957]).

In a geological classification system, the composition, texture, and structure of a sediment and the fossils it contains are related to origin, aiming at the reconstruction of the geological history: generic formation (extrabasinal—terrigenous, clastic sediments, and intrabasinal—in-place formed biogenous and chemical or hydrogenous sediments), agents of deposition (transportation of sediments by the action of waves, currents, tides, gravity, ice, winds, and organisms), and sedimentary environments (e.g., nearshore, shallow sea, continental slope, pelagic, and glaciomarine environments). In the first classification system, the clastic sediments are subdivided into textural classes (i.e., gravel, sand, silt, clay), and the biogenic and

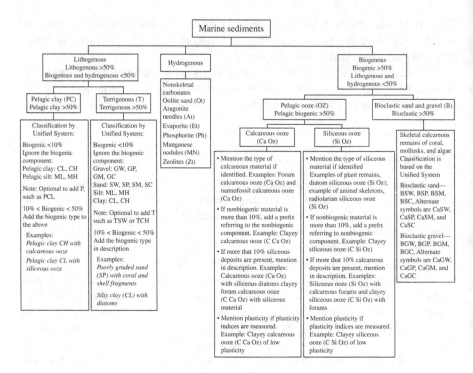

FIGURE 7.10 Chart for classification of marine sediments.

(From Noorany [1989]. With permission from ASCE)

chemical sediments are classified according to their major chemical or mineralogical constituents.

A principal objective of engineering classifications is to allow a reasonable assessment of the fundamental properties of a given soil by comparison with another that has similar behavior. This comparison allows the reduction or the better direction of complex testing procedures that are employed to evaluate them. Therefore, a classification system must employ simple, easily obtained criteria, such as those evaluated by simple index tests. It will lose its value if the tests become more complicated than the tests to measure directly the fundamental properties. Casagrande (1948) provided some guidelines for classification systems, which are as follows: (1) properties in undisturbed condition, (2) indication of properties, (3) applicability in both visual and laboratory references, and (4) simple system of notation. The methods of classifying soils for engineering purposes are based on either the properties of raw materials of soils, or on properties of the soil in its undisturbed condition. Within these two basic methods, Casagrande (1948) in Figure 7.11 outlined the various techniques available. Under properties of raw materials of soils, there are: (1) textural soil classifications and (2) road subgrade classifications (such as American Association of State Highway and Transportation Officials and airfield subgrade classifications).

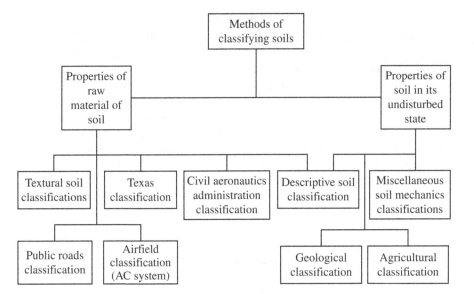

FIGURE 7.11 Engineering methods of classifying soils.

(Adapted from Casagrande [1948]. With permission of ASCE)

7.7.3 CLASSIFICATION OF SILICICLASTIC SEDIMENTS

Soil classification deals with the categorization of soils based on distinguishing characteristics as well as potential criteria that dictates choices in use. Soil classifications arise from the need for (1) rational, yet convenient characterization of soils; (2) their correlation (differentiation, comparison, and contrast), and (3) methods for effective communization based on a uniformity of criteria and nomenclature.

7.7.3.1 Types of Classification Systems

Soil characteristics most used in engineering classification systems are particle size and consistency (Atterberg limits). These simple tests reflect the soil's physical properties, composition, moisture content, and so on. The information obtained by these tests enables an individual to assess the quality of a soil for construction or as a foundation. Several systems are in use in the United States and throughout the world, which have application in the marine environment. These are the USCS, which was originally proposed by A. Casagrande in 1942 and was later revised and adopted in 1952 by the U.S. Bureau of Reclamation and U.S. Army Corps of Engineers and later by the ASTM D2487. In one form or another the unified system is used for virtually all geotechnical work throughout the world. In these systems soils are classified by the dominant particle size and organic content into gravels (G), sands (S), silts (M), clays (C), organic sediments (O), and peats (Pt). Coarse-grained or granular soils (G and S) are further subdivided by grading, and fine-grained or cohesive soils (M, C, and O) by liquid limit and plasticity index, into 15 groups, which have distinct engineering properties.

7.7.3.2   Classification by End Member Concept

One of the most useful methodologies of classifying sediments is by using the end member concept (i.e., sand, silt, clay) (Shepard, 1954). Sediments containing three constituents can be classified in a triangular diagram in which each apex represents 100% of one of the three constituents (i.e., sand, silt, and clay). More general classification schemes based primarily on sediment composition but also on origin have also been presented. Lisitzin (1972) proposed a system in which he divided sediments into terrigenous, biogenic, chemogenic, volcanogenic, and polygenic (red clay). Berger and von Rad (1972) followed this effort with a classification system for the predominate deep-sea sediments: pelagic and hemipelagic materials. Hemipelagic sediments are not deposited as slowly as pelagic sediments. They are deposited on continental shelves and rises, and ordinarily accumulate too rapidly to react chemically with seawater. Berger's classification system does not extend to a variety of other sediments.

This methodology can also be used to classify a soil comprised of mixed sediments, non-clastic sediments, environment-dominated sediments or other sediment types, provided they possess a common parameter that can be expressed quantitatively (e.g., percentage). Four-component sediment systems can be plotted in three dimensions within the faces of a tetrahedron.

7.7.3.3   In-Situ Classification Methods

Classification of sediments, evaluation of their subsurface stratigraphy, and the lateral extent (soil profiling) can be accomplished by the interpretation of results obtained by in-situ penetration testing. The charts resulting from penetration tests are also used to classify soil types based on soil behavior. They are not grain size classification charts. For this reason, some overlap exists between different types of soils that have entirely different consistencies, for example, loose sands and highly over-consolidated clays. Therefore, the interpretation of penetration testing data must be supported by intelligent assessment of the data, and additional data provided by use of specific tools and experience.

The advantages of a cone penetration test (CPT) are speed and economy, applicability in almost all soil types, weak rocks, and provision of detailed, precise, and continuous data which are better suited to many ordinary soil engineering design problems. However, problems associated with in situ methods include lack of soil samples for both visual inspection and laboratory testing, depth penetration is sometimes severely limited, and the electronic control and data retrieval equipment is expensive and may be easily damaged.

7.7.4   CLASSIFICATION OF FINE-GRAINED SEDIMENTS

The classification of fine-grained sediments can be determined by particle size and consistency (Atterberg Limits, ASTM D4318). These simple tests reflect the soil's physical properties, composition, and moisture content. To determine particle size of a sample, two procedures are typically employed. These methods are the hydrometer test (ASTM D2216) and the pipette method (i.e., fall velocity). The hydrometer method involves dispersing a soil sample in a known volume of water and monitoring the change in the solutions specific gravity using a hydrometer as a function of time. A second method utilized to define a particles diameter is by its fall velocity. A sediment particles diameter is modeled as a sphere having the same density and fall velocity. This definition has the advantage of relating a particle diameter to its fluid behavior, which is the normal reason for needing to determine the diameter. However, a settling tube analysis is not as reproducible as a sieve analysis. Standardized testing procedures have not yet been agreed upon. Other diameter definitions that have been occasionally used include the nominal diameter, the diameter of a sphere having the same volume as the particle; and the axial diameter (i.e., the length of one of the grain's principal axes), or some combination of these axes.

For nearly spherical sand particles there is little difference in these definitions. When reporting the results of an analysis, it is always appropriate to define the diameter (or describe the measurement procedure), particularly if the sieve diameter is not being used.

There is a need to characterize an appropriate diameter for an aggregation of particles, rather than the diameter of a single particle. Even the best-sorted natural sediments have a range of grain sizes.

Sediment samples found in nature typically have a few large particles covering a wide range of diameters and many small particles within a small range of diameters. That is, most natural sediment samples normally have a highly skewed distribution. However, if size classes are based upon a logarithmic (power of 2) scale and each size fraction are considered by weight, then typical sediment samples will have a normal distribution.

A sediment sample plotted as a percentage of the total weight of the sample being sieved versus the particle size approximates a straight line on a log-normal graph. This line is known as the log-normal distribution. A distribution of this type can be described using standard statistical parameters. The sediment size data comes from the weight of that accumulates on each sieve in a nest of graduated sieves. This can be plotted on semi-log paper.

7.8 PHYSICAL PROPERTIES

7.8.1 INTRODUCTION

Bulk properties are specifically related to the mechanical and hydraulic behavior of a material. To present this topic a number of terms need to be defined. These terms are (1) bulk density, (2) density of solids, (3) dry density/dry unit weight, (4) porosity, (5) void ratio, and (6) total density/total unit weight.

7.8.1.1 Density

7.8.1.1.1 Total Density

The total wet density is usually referred to as the bulk density or total density (ρ_T). It is defined as the total wet sample mass (M_t) divided by the total wet sample volume (V_t), or

$$\rho_T = \frac{M_T}{V_T}.$$ (7.10)

Bulk density is essential for evaluation of sediment stress (consolidation history), correlation of repeatedly drilled cores, correlation with downhole logs, calculation of geophysical seismograms, and so on. It can be an excellent proxy for environmental changes, because it reflects compositional changes in the sediment; these changes may be related to warm and cold cycles, changes in productivity or preservation, or the control of changing sea level on sediment input. Soil is composed ideally of three phases (i.e., gas, water, and solids) as shown in Figure 7.12.

A review of Figure 7.12 shows that M_T and V_T used for bulk density are the total mass and volume of the three phases, solid (M_s, V_s), liquid (M_w, V_w), and gas (M_g, V_g). However, gas if present in situ, escapes as the sediment cores are taken, cut into sections, and possibly split into halves for sedimentological description. Therefore, the fluid phase is assumed equal to the pore water phase, neglecting the unrecovered gas phase. The equations for total wet sample mass (M_t) and total wet sample volume (V_t) is as follows:

$$M_T = M_s + M_w. \tag{7.11}$$

$$V_T = V_S + V_W. \tag{7.12}$$

It is furthermore assumed that sediments sampled under water are saturated. For samples taken on the beach, this is not true. It is normally not possible to evaluate the degree of saturation because of problems accounting for the gas phase using routine techniques.

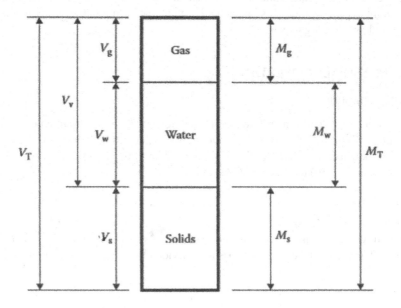

FIGURE 7.12   Sediment phase relation diagram.

TABLE 7.10
Typical Engineering Values

Material		Dry Bulk Density (kg/m³)	Saturated Bulk Density (kg/m³)	Difference (kg/m³)
Uniform sand				
	Loose	1430	1890	460
	Dense	1750	2090	340
Mixed sand				
	Loose	1590	1990	400
	Dense	1860	2160	300
Clay				
	Stiff glacial		2070	
	Soft, very organic		1430	

Source: Terzaghi and Peck (1967).

For uniform solid and pore fluid densities, bulk density is therefore simply a function of the water content. Table 7.10 lists typical bulk densities for several sediments in the coastal zone.

In comparison, Table 7.11 lists mean porosity and saturated bulk density of naturally occurring soils. A review shows that the density of naturally compacted sand on the seabed and shoreline is between the typical engineering values for dry and saturated sands recommended by Terzaghi and Peck (1967). The saturated bulk density for soft very organic clays in Table 7.10 agrees with the value for natural sand-silt-clay sediments reported by Manger (1966).

Comparison of the two columns of data in Table 7.12 gives an idea of the consolidation to be expected from settling (i.e., difference between dense and loose), and a minimum estimate of newly (i.e., dense) placed dry material.

TABLE 7.11
Natural Surface Soils

Material	Mean Porosity (%)	Saturated Bulk Density (kg/m³)	Location
Sand	38.9	1930	Cape May sand spits
Loess	61.2	1610	Idaho
Fine sand	46.2	1930	CA seafloor
Very fine sand	47.7	1920	CA seafloor
Sand-silt-clay	74.7	1440	CA seafloor

Source: Daly et al. (1966).

TABLE 7.12

Typical Data Laboratory Soils

Material	Loose Dry Bulk Density (kg/m³)	Dense Dry Bulk Density (kg/m³) Tapped	Difference (kg/m³)
Gravelly sand	1660	1770	110
Sandy soil	1440	1560	120
Dune sand	1610	1760	150
Loess	990	1090	100
Peat	270	320	50
Muck	800	850	50

Source: Johnson and Olhoeft (1984).

7.8.1.1.2 Density of Solids (Particles or Grain Density)

Density of solids (particle or grain density) is defined as the mass of the solids M_s (mineral grains) divided by their volume V_s:

$$\rho_s = \frac{M_s}{V_s}. \tag{7.13}$$

The density of solids (grain density) is directly determined either by using a precision balance and a pycnometer or by calculation from bulk density and the water content. The procedure assumes that oven drying has driven all pore fluids from the sample, leaving the precipitated pore water salts with the solid matter. The specific gravity of solids can then be determined as follows:

$$G_S = \frac{\rho_s}{\rho_w}. \tag{7.14}$$

Typical solid densities are 2.65 g/cm³ for quartz and 2.72 g/cm³ for carbonates. The total wet volume (V_t) can be determined using either a ring of known volume or a pycnometer. The bulk density can then be calculated directly (M_t/V_t). The density of solids (grain density) can then be computed indirectly by subtracting the mass of the pore water (M_w) from the total mass (M_t) divided by the total volume (V_t) minus the volume of water (V_w).

$$\rho_S = \frac{M_t - M_w}{V_t - V_w}, \tag{7.15}$$

where V_w is calculated based on Figure 7.12, which is as follows:

$$V_w = \frac{M_w}{\rho_w}, \tag{7.16}$$

where $\rho_w = \dfrac{M_w}{V_w}$, M_w mass of water, V_w volume of water.

TABLE 7.13
Typical Grain Densities of Common Sediments

Location	Packing Arrangement	Material	Condition	Porosity (%)	Reference
	Cubic	Uniform spheres	Loosest	0.476	
	Rhombohedral	Uniform spheres	Densest	0.260	
		Natural sand		$0.25 < n < 0.50$	Blatt et al. (1980)
Lab		Poorly graded sand		0.424	Blatt et al. (1980) and Terzaghi and Peck (1967)
Lab		Well-graded sand		0.279	Blatt et al. (1980) and Terzaghi and Peck (1967)
Scripps Canyon, CA		Fine sand		0.42 beach; 0.40 compacted by vibration; 0.50 offshore	Chamberlain (1960) and Dill (1964)
Scripps Canyon, CA		Micaceous sand		0.62	Chamberlain (1960) and Dill (1964)

Source: U.S. Army Corps of Engineers, *Shore Protection Manual* (2008). Courtesy of U.S. Army.

If the dry volume V_d determined using a pycnometer is measured, then the density of solids ρ_s (i.e., grain density) is calculated directly (M_s/V_s) and the bulk density (ρ) is derived indirectly by adding the pore water mass (M_w) and volume (V_w) to the solid mass and volume.

$$\rho = \frac{M_s + M_w}{V_s + V_w}, \tag{7.17}$$

where M_w, V_w, and ρ_w are calculated using the phase diagram in Figure 7.12 or Equation (7.16), respectively. Typical grain densities of minerals in the coastal zone are presented in Table 7.13.

7.8.1.1.3 Dry Density/Dry Unit Weight
Dry density is used to estimate mass accumulation, or alternatively, to judge the degree of compaction. It is defined as the ratio of mass of solids M_s to the total volume V_t:

$$\rho_d = \frac{M_s}{V_t}. \tag{7.18}$$

It can also be expressed in terms of water content and bulk density calculated above:

$$\rho_d = \frac{\rho}{(1+w)}, \tag{7.19}$$

where water content (w) is defined as follows:

$$w = \frac{M_w}{M_s},\tag{7.20}$$

where:

M_w—mass of water
M_s—mass of solids

These relationships show that dry density is an important parameter when materials of significantly different solid density, such as carbonate and organic matter, were accumulated at different rates with time or depth (Equation 7.21). For uniform density of solids, dry density is simply a function of the water content. The mass density of various shoreline rocks is presented in Table 7.14.

TABLE 7.14
Average Mass Densities of Coastal Rocks

Number	Rock Type	Mass Density (kg/m³)
1	Basalt	2.74
2	Dolerite-Diabase	2.89
3	Granite	2.66
4	Sandstone	2.22
5	Dolomite	2.77
6	Limestone	2.54

Source: U.S. Army Corps of Engineers, *Shore Protection Manual* (2008). Courtesy of U.S. Army.

$$\gamma = \rho g,\tag{7.21}$$

where:

ρ is the denotes density
g is the acceleration due to gravity

Then the dry unit weight is as follows:

$$\gamma_d = \rho_d g.\tag{7.22}$$

7.8.1.2 Porosity

Porosity (n) is defined as the ratio of volume of pore space or voids V_v to total volume V_t times 100%.

$$n = \frac{V_v}{V_t}100\%.\tag{7.23}$$

The volume of voids is defined as the volume of the pore fluids, including the volume of air or other gases V_g:

$$V_v = V_w + V_g. \tag{7.24}$$

Marine and coastal specimens are assumed saturated because the volume of gas V_g cannot normally be measured and is therefore neglected. The volume of voids therefore equals the volume of pore water.

This relationship can be expressed in terms of pore water content (w, in %) and bulk density (ρ) as determined above, and the pore water density ρ_w.

$$n = \frac{w\rho}{\rho_w (1+w)}. \tag{7.25}$$

For uniform solid and pore fluid densities, porosity is therefore simply a function of the water content.

A measure of how tightly particles are packed together is the porosity. Therefore, porosity is not a constant for a given sediment. As a particle settles to the bed, the effect of gravity is less than the effect of the fluid stresses over the bed. Lower compaction is the result of this combination of forces. Therefore, the particles have less opportunity to move and find a position of maximum stability. Thus, particles in the surf zone are typically compacted to near their maximum compaction. This is not the case in many quiet estuaries. The typical variation of porosities is presented in Table 7.15.

TABLE 7.15
Porosity Variation

Location	Packing Arrangement	Material	Condition	Porosity (%)	Ref:
	Cubic	Uniform Spheres	Loosest	0.476	
	Rhombohedral	Uniform Spheres	densest	0.260	
		Natural sand		$0.25 < n < 0.50$	Blatt et al., 1980
Lab		Poorly Graded Sand		0.424	Blatt et al., 1980; Terzaghi and Peck. 1967
Lab		Well graded sand		0.279	Blatt et al., 1980; Terzaghi and Peck. 1967
Scripps Canyon, Ca		Fine Sand		0.42 beach; 0.40 compacted by Vibration; 0.50 offshore	Chamberlain 1960; Dill, 1964
Scripp Canyon, Ca		Micaceous sand		0.62	Chamberlain 1960; Dill, 1964

In natural sands, compaction is essentially independent of particle size. However, the compaction is complicated by both irregular shapes and nonuniform size of particle. An increase in nonuniformity of particles sizes (i.e., well graded) increases in general the compaction (decreases the porosity). This occurs because small particles can fit into the pore spaces between the large particles. In engineering terms, porosity decreases as the sand becomes more poorly (i.e., uniform) graded. In geology terms the sand is now called well graded. The compaction tends to decrease as the regularity of the particle shapes increase.

Good porosity data are not often available. The standard assumption in longshore transport computations is that sand has a porosity of 0.40, although there are likely to be significant variations from that figure, as discussed by Galvin (1979).

7.8.1.3 Void Ratio

Void ratio (e) is defined as the volume of voids (V_V) to the volume of solids (V_S),

$$e = \frac{V_V}{V_s}.$$ (7.26)

Therefore

$$n = \frac{e}{1+e}.$$ (7.27)

Void ratio (e) can also be calculated from water content (w), the density of solids (ρ_s) derived above, and the density of pore water (ρ_w). Therefore for a saturated condition the void ratio can be calculated as given below:

$$e = \frac{w\rho_s}{\rho_w}.$$ (7.28)

This can be altered by defining saturation (S) as follows:

$$S = \frac{V_w}{V_v},$$ (7.29)

where V_w is the volume of water.

Then rewriting Equation (7.28) gives the following relation:

$$Se = wG_s.$$ (7.30)

For uniform solid and pore fluid densities under saturated conditions, void ratio is simply a function of the water content.

Another parameter that is utilized to describe packing along with porosity and void ratio is the specific volume (v). The specific volume (v) is defined as the volume of the soil sample containing a unit volume of soil grains as shown in Figure 7.13.

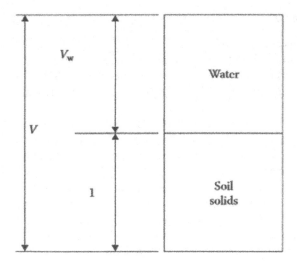

FIGURE 7.13 Specific volume.

Since $V_v = eV_s$, $V_v = V_w$ for saturated sediments, therefore:

$$V_T = 1 + V_w = 1 + V_V,$$ (7.31)

$$V_T = 1 + eV_s,$$ (7.32)

$$V_T = 1 + e.$$ (7.33)

7.8.1.4 Total Density/Total Unit Weight

The total density of the sediment (ρ) is given by Equation (7.34).

$$\rho = \left(\frac{G_s + Se}{1 + e}\right)\rho_w.$$ (7.34)

$$\rho = \left(\frac{1 + w}{1 + e}\right)G_s\rho_w.$$ (7.35)

The corresponding total unit weight (γ_T) is as follows:

$$\gamma_T = \left(\frac{G_s + Se}{1 + e}\right)\gamma_w.$$ (7.36)

$$\gamma_T = \left(\frac{1 + w}{1 + e}\right)G_s\gamma_w.$$ (7.37)

7.8.2 Permeability

The void spaces between soil grains allow liquids to flow through them. A measure of this ability of a soil to transmit liquids is its permeability. Factors affecting permeability are: (1) the effective grain size or effective pore size; (2) shapes of voids and flowpaths through the soil pores—tortuosity; (3) saturation; and (4) viscosity of permeant.

Darcy (1856) proposed an equation for calculating the velocity of flow of water through a sediment. This equation was originally developed for sandy materials, which assumes a linear relationship between the average surficial velocity of the fluid and the hydraulic gradient, with the proportionality constant or coefficient of permeability, k, being unique for a given sediment at a particular porosity, temperature, and pressure. Other, more complex relationships have been suggested in attempts to account for actual flow conditions through fine-grained soils (Michaels and Lin, 1954; Mitchell, 1976; Olsen, 1965). A number of these relations depend upon various functions involving void ratio (refer to Figure 7.14).

This Darcy relationship for velocity is presented in the following equation:

$$v = ki, \tag{7.38}$$

where:

v is the velocity
k is the coefficient of permeability
i is the hydraulic gradient (dimensionless)

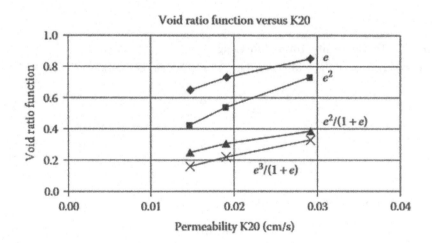

FIGURE 7.14   Permeability versus various void ratio functions for a rounded sand.

(Chaney and Almagor, 2016. Republished with permission of Taylor and Francis)

The hydraulic gradient is defined as follows:

$$i = \frac{\Delta h}{L},$$

(7.39)

where:

Δh is the piezometric head difference between the sections AA and BB (Figure 7.15)
L is the distance between sections AA and BB

Water migration due to hydraulic gradients in marine sediments can be induced by natural processes such as sediment accumulation proposed by Gibson (1958) or by placement of structures on or in the seafloor. Water movement through clays is a very complex phenomenon because of the physicochemical activity of the particles and the effect of the surrounding double layers on the clay particles.

Silva et al. (1981) reported results of low-gradient permeability testing of fine-grained marine sediments using a modified back-pressured consolidation system for direct permeability measurements. They ran their experiments by first consolidating a sample under a given effective stress. Once an equilibrium state was reached, a constant head permeability test was then conducted. These tests were conducted at varying gradients to study the validity of Darcy's equation and the occurrence of a threshold gradient. They found that most undisturbed samples (illite and smectite clays) did not show evidence of a threshold gradient. A threshold gradient was shown to become more probable as the sediment becomes denser. They did find that some artificially sedimented samples of illitic clay indicated threshold gradients of less than 5.

Results of tests on five different types of deep-sea sediment are shown in Figure 7.16. These tests were performed on a combination of undisturbed and artificially sedimented marine sediments. A review of this figure shows that the range of permeability is a function of material type and void ratio.

One source of energy dissipation for waves traveling in shallow water is the flow into and out of the sediment bed (Packwood and Peregrine, 1980; Reid and Kajiura, 1957).

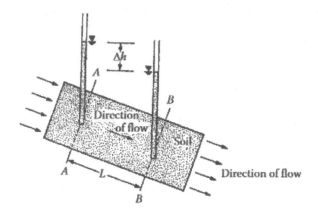

FIGURE 7.15   Schematic definition of hydraulic gradient.

(Chaney and Almagor, 2016. Republished with permission of Taylor & Francis)

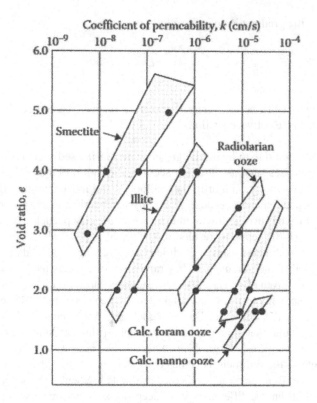

FIGURE 7.16 Comparative void ratio versus coefficient of permeability plots showing ranges of measurements made for five general deep-sea sediment types.

(Silva et al., 1981)

The steepness of the foreshore is determined primarily by the sediment permeability. During the wave uprush in the swash zone sediment is carried shoreward. The water returning to the sea above and through the bed and on the surface is controlled by the permeability in the swash zone.

The surface backrush will transport sediment seaward. This action decreases the equilibrium foreshore slope (McLean and Kirk, 1969; Packwood, 1983; Savage, 1958). Seaward of the breaker zone recent studies have indicated small amounts of wave-induced flow into and out of the seabed have a significant effect on the bottom boundary layer and the resulting sediment transport (Conley and Inman, 1992).

7.8.3 STRENGTH OF SEDIMENTS

In the following the strength of marine clay will be first discussed followed by sands. The idealized shear strength profiles for homogeneous marine deposits that are (1) NC clay—normally consolidated, (2) UC clay—under consolidated, and (3) OC clay—over consolidated are presented in Figure 7.17. A review of Figure 7.17a shows that the NC clay profile exhibits a linear shear strength behavior starting at zero at

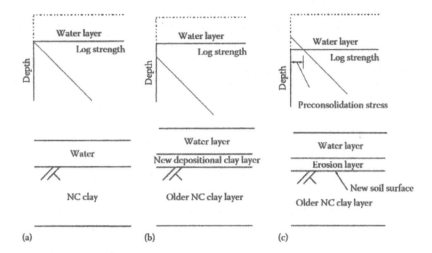

FIGURE 7.17 Idealized strength profile in homogeneous clay: (a) normally consolidated clay (NC) strength profile; (b) under consolidated clay (UC) strength profile; (c) over consolidated clay (OC) strength profile.

(Chaney and Almagor, 2016. Republished with permission of Taylor & Francis)

the water/sediment interface and increasing with depth. Figure 7.17b depicts a UC clay profile where a new clay layer overlays an older clay layer that has not reached equilibrium under its previous loading. This new clay layer results in an increase in excess pore pressure in the older layer. The resulting shear strength in the older clay layer is therefore decreased. For an OC clay (Figure 7.17c), the clay layer has experienced a larger load in its geologic past than exists at present. The clay layer remembers this past loading in the form of a cohesion and the resulting shear strength versus depth is increased.

The strength of sediments in the above is a function of its structure (i.e., assemblage of grains), moisture content, and the properties of the mineral grains forming the solid phase of the system. In this section, the influence of these properties is discussed. The strength and shearing resistance of soils (sediments) are the most important characteristics for engineering design or consideration of hazards. In the following, the theoretical background to shear strength will be discussed using empirical, Mohr–Coulomb, and normalized behavior approaches to estimating the shear strength of soils.

7.8.3.1 Mohr–Coulomb Approach

For an NC clay or a sandy clay in the ground, a relation between undrained shear strength (S_u), effective vertical stress (σ'_{vo}), lateral earth pressure coefficient at rest (K_o), and the effective friction angle (φ') can be derived as follows: Consider a soil element at A in Figure 7.18.

The major and minor effective principal stresses at A can be given by σ'_{vo} and $K_o \sigma'_{vo}$, respectively. K_o is the coefficient of earth pressure at rest. Let the soil element be subjected to an unconsolidated undrained (UU) triaxial test. The total and effective stress Mohr's circles for this test, at failure, are shown in Figure 7.19.

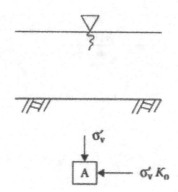

FIGURE 7.18   In-situ condition existing on element A.

(Chaney and Almagor, 2016. Republished with permission of Taylor & Francis)

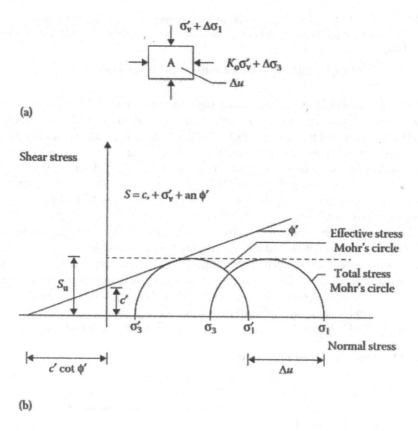

FIGURE 7.19   Total and effective Mohr's circles. For a specimen exhibiting positive pore water pressure: (a) element A at failure; (b) total and effective Mohr's circles for element A at failure.

(Chaney and Almagor, 2016. Republished with permission of Taylor & Francis)

A review of this figure shows that at failure the total major principal stress is $\sigma_1 = \sigma_v' + \Delta\sigma_1$; the total minor principal stress is $\sigma_3 = K_o\sigma_v' + \Delta\sigma_3$ and the excess pore water pressure is Δu. The corresponding effective major and minor principal stresses can be given by $\sigma_1' = \sigma_1 - \Delta u$ and $\sigma_3' = \sigma_3 - \Delta u$, respectively. Using Figure 7.18, the following equations:

$$\sin\phi' = \frac{S_u}{c'\cot\phi' + \left(\dfrac{\sigma_1' + \sigma_3'}{2}\right)}$$

$$S_u = c'\cot\phi + \left(\frac{\sigma_1' + \sigma_3'}{2}\right)\sin\phi' \tag{7.40}$$

$$S_u = c'\cot\phi' + \left(\frac{\sigma_1' + \sigma_3'}{2} - \sigma_3'\right)\sin\phi' + \sigma_3'\sin\phi'.$$

$$S_u = \left(\frac{\sigma_1' + \sigma_3'}{2} - \sigma_3'\right) = \left(\frac{\sigma_1' + \sigma_3'}{2}\right). \tag{7.41}$$

$$S_u = c'\cot\phi' + S_u\sin\phi' + \sigma_3'\sin\phi'$$
$$S_u(1 - \sin\phi') = c'\cot\phi' + \sigma_3'\sin\phi' \tag{7.42}$$
$$\sigma_3' = \sigma_3 - \Delta u = K_o\sigma_v' + \Delta\sigma_3 - \Delta u.$$

From Skempton (1954),

$$\Delta u = B\Delta\sigma_3 + A_f(\Delta\sigma_1 - \Delta\sigma_3). \tag{7.43}$$

For saturated clay, $B = 1$ and substituting the above equation into Equation (7.42):

$$\sigma_3' = K_o\sigma_v' + \Delta\sigma_3 - [\Delta\sigma_3 + A_f(\Delta\sigma_1 - \Delta\sigma_3)]$$
$$= K_o\sigma_v' + A_f(\Delta\sigma_1 - \Delta\sigma_3)$$
$$S_u = \frac{\sigma_1 - \sigma_3}{2} = \frac{(\sigma_{vo}' + \Delta\sigma_1) - (K_o\sigma_v' + \Delta\sigma_3)}{2} \tag{7.44}$$
$$2S_u = (\Delta\sigma_1 - \Delta\sigma_3) + (\sigma_{vo}' + K_o\sigma_{vo}')$$
$$(\Delta\sigma_1 - \Delta\sigma_3) = 2S_u - (\sigma_v' - K_o\sigma_v').$$

Substituting Equation (7.42) into Equation (7.44) gives

$$\sigma_3' = K_o\sigma_{vo}' - 2S_u A_f + A_f\sigma_v'(1 - K_o)$$
$$S_u = \frac{c'\cot\phi' + \sigma_v'\sin\phi'[K_o + A_f(1 - K_o)]}{1 + (2A_f - 1)\sin\phi'}. \tag{7.45}$$

For NC clays, $C = 0$; therefore,

$$\frac{S_u}{\sigma'_{vo}} = \frac{\sin\phi'[K_o + A_f(1 - K_o)]}{1 + (2A_f - 1)\sin\phi'}. \tag{7.46}$$

where:

σ'_v is the vertical effective stress.
K_o is the earth pressure coefficient at rest.
A_f is the pore pressure parameter A at failure.

For stress conditions associated with no lateral yielding, it might be assumed to exist. During deposition either horizontally or on a gentle inclination, K_o may be expressed empirically.

$$K_o = 1 - \sin\phi'. \tag{7.47}$$

7.8.3.2   Empirical Approach

For stable clays, φ' varies between 20 and 35 degrees. Stable loose silts and sands typically have values of φ' between 28 and 34 degrees. In addition, a correlation between φ' and plasticity index has been given by Bjerrum and Simons (1960).

Large deformations in soils containing a clay content greater than approximately 35% induce preferred orientation of the clay particles in the shear zone and cause a reduction of φ' (Skempton, 1964). Angles of shearing resistance as low as 10 degrees are not uncommon in clays that have been subject to large strains. Few data giving strength parameters in terms of effective stress are available for present-day marine sediments.

When a fully saturated soil is sheared under undrained conditions and the results are interpreted in terms of total stresses, the material behaves as though it is purely cohesive. This holds for saturated sand as well as for clays (Bishop and Eldin, 1950).

7.8.3.3   Normalized Behavior Approach

For any particular NC soil, the ratio

$$\frac{S_u}{\sigma'_v} \tag{7.48}$$

is relatively constant and indicates that the undrained strength increases with depth. The behavior of soft clay based on laboratory experimentation was modeled as an ideal soil. The work was done on remolded soil but was found to model a number of natural soils. The ideal behavior consisted of the following:

- Strains would occur only if effective stress condition changed.
- If effective stress remained constant, no strains occurred.

- Secondary consolidation and creep were neglected.
- Results of consolidation tests on NC sediments when plotted as the log of the effective stress (σ'_v) versus void ratio (e) resulted in a straight line.
- Points representing OC states and having equal values of OCRs (s'_m/s'_v) plotted as lines parallel to the normally consolidation line, where p'_m is the maximum past effective stress and p' is the existing effective stress.
- Results of identical types of strength tests on an NC soil expressed as shear strength versus volumetric strain were found to plot as a straight line parallel to the normal consolidation line. The ratio of strength to consolidation pressure was constant for each particular type of test, $S_u/\sigma'_v = C2$, where C2 is a constant dependent on the value of the OCR and the type of test.
- The area under the curve of log σ'_v versus volumetric strain represents the net mechanical work W done on the structure of the soil during consolidation and swelling (Kenney and Folkes, 1979).

A soil that exhibits normalized behavior (i.e., the ratio of S_u/σ'_v is constant) is characterized by a linear curve between the water content (or void ratio) and the log σ'_v as shown in Figure 7.20a. By contrast, a soil that does not exhibit normalized behavior (the ratio S_u/σ'_v is not constant) in a manner shown in Figure 7.20b. Materials that do not exhibit normalized behavior are quick or naturally cemented clays with a high degree of structure.

A design procedure for use with clay deposits exhibiting normalized behavior was developed by Ladd and Foote (1974) and later refined by Ladd (1991). The procedure is called SHANSEP (stress history and normalized soil engineering properties). The method, which was developed for soft clays, attempts to normalize the undrained shear strength (S_u) with respect to the in-situ vertical effective stress (s'_v). The approach is based on the fact that soft clays with the same OCR, but different consolidation stresses, display similar strength and stress–strain characteristics. Thus,

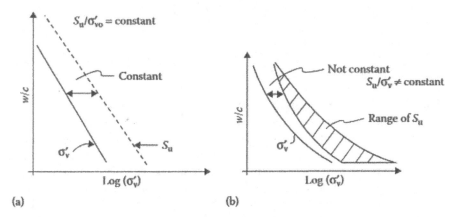

FIGURE 7.20 Normalized behavior of soil: (a) Material exhibiting normalized behavior; (b) material not exhibiting normalized behavior. *Note*: w/c is the water content.

(Chaney and Almagor, 2016. Republished with permission of Taylor & Francis)

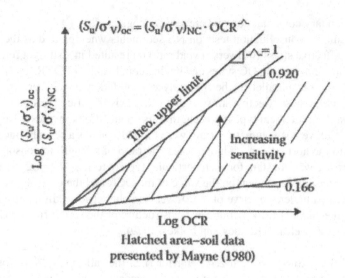

Hatched area–soil data
presented by Mayne (1980)

FIGURE 7.21   Schematic representation of $(Su/\sigma'_c)OC/(Su/\sigma'_c)$ NC versus OCR. (Chaney and Fang, 1986)

the normalization procedure provides a dimensionless constant, linking the existing effective vertical stress to the undrained shear strength. Using the above laboratory testing approach, normalized parameters are obtained and then applied to the stress history as reflected by the materials' OCR and its preconsolidation pressure (σ'_m) of the cohesive deposits to derive strength–deformation properties for use in analysis. Application of this methodology utilizes the following equation:

$$\left(\frac{S_u}{\sigma'_v}\right)_{OC} = \left(\frac{S_u}{\sigma'_v}\right)_{NC} OCR^\wedge . \tag{7.49}$$

Mayne showed that the parameter (Λ) varies from 0.166 to 0.920 as the materials' sensitivity increases (see Figure 7.21).

7.8.3.4   Static Shear Strength Behavior of Sands and Clays

Typical profiles of S_u versus depth (0 to 30 m) for various marine soils are shown in Figure 7.22. The strength profiles presented are for: (1) hemipelagic, terrigenous silty clay in the Santa Barbara Channel; (2) turbidites; (3) calcareous ooze; (4) pelagic clays, and (5) siliceous ooze. A similar attempt at summarizing the distribution of shear strength in the upper approximately 1 m of sediment over the North Pacific Ocean basins was presented by Keller (1968). The shear strength values that Keller presented were made using either a laboratory miniature vane or the UU triaxial test. A review indicates that large areas associated with red clay have shear strengths <3.5 kPa. In coastal areas, the strength values <3.5 kPa are found, although small variations in the environment result in major changes.

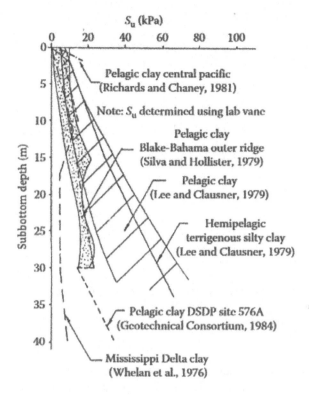

FIGURE 7.22   Typical clay strength profiles.

(Chaney and Fang, 1986)

7.8.3.5   Cyclic Behavior of Granular Materials

If a relatively loose (relative density <70%), saturated sandy soil is subjected to cyclic loads it tends to compact and decrease in volume. If drainage is prohibited during the cyclic loading there is a subsequent buildup in pore water pressure until it is equal to the overburden pressure. The effective stress becomes zero and the soil loses its shear strength and develops a condition where it no longer can support shear loads (liquefied condition). The factors that influence cyclic loading-induced liquefaction are (1) soil type, (2) relative density or void ratio, (3) initial confining pressure, (4) nature of cyclic loading (magnitude and duration), and (5) level of saturation (Chaney, 1978).

A typical test record of a cyclic loading test on a sand material shows that as the cyclic loading is occurring there is a corresponding buildup in the pore water pressure and a subsequent axial strain. If this condition is of a large extent, and the pore water pressure is not relieved, a lateral movement may result.

The corrected stress ratio (i.e., cyclic stress ratio) as a function of the number of cycles of loading is shown for Monterey No. 0 sand at four relative densities in Figure 7.23. A review of Figure 7.23 shows that for increasing relative densities the number of load cycles to cause initial liquefaction increases.

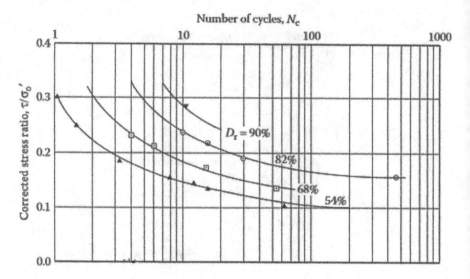

FIGURE 7.23 Corrected cyclic stress ratio and number of load cycles to cause initial lique-faction of sand at different initial relative densities.

(DeAlba et al., 1976)

Liquefaction is fundamentally controlled more by shear strain than by shear stress. This results from the generation of pore pressure due to the shear strain breaking down the soil structure and a corresponding tendency to densify. There is a level of shear strain, or threshold shear strain below which no pore pressure is generated. Using the volumetric strain (densification) that would occur if drainage were permitted and the slope of the rebound curve (E) the induced pore pressure can be determined as shown in the following equation.

$$\Delta u = E_r \Delta \varepsilon_{rd}. \tag{7.50}$$

Martin et al. (1975) have given procedures to evaluate these two parameters from the results of static rebound tests in a consolidation ring and cyclic load tests on dry sand, respectively. Finn and Bhatia (1981) have also reported good agreement between predicted and measured values using the proposed method.

A relation has been presented by Martin et al. (1975) that relates the increase in residual pore water (u) for each load cycle to the rebound tangent modulus (E_r), porosity (n), bulk modulus of an air/water mixture (K_{aw}), and the volumetric strain per cycle ($\Delta \varepsilon_{vd}$), which is given by Equation (7.51).

$$\frac{\Delta u}{\Delta \varepsilon_{vd}} = \frac{1}{\left[\left(\frac{1}{E_r}\right) + \left(\frac{n}{K_{aw}}\right)\right]}. \tag{7.51}$$

Assuming various values of and substituting a value of K_{aw} at atmospheric pressure is a qualitative measure of the effect of $\Delta \varepsilon_{vd}$ on pore pressure increment (Figure 7.24).

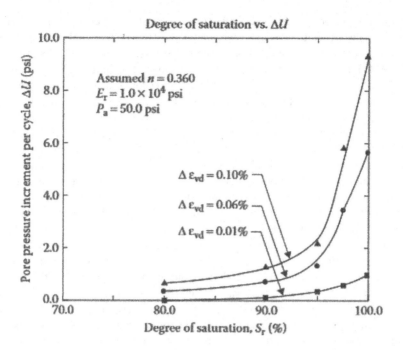

FIGURE 7.24 Change in pore pressure per load cycle as a function of degree of saturation. (Chaney, 1978)

A review shows that for saturation ($S = 100\%$) the residual pore pressure per cycle does not start to increase until a volumetric strain per cycle $\Delta\varepsilon_{vd}$ of $10^{-2}\%$.

7.8.3.5.1 Mechanism and Implications of Liquefaction Phenomena

The basic cause of liquefaction in saturated cohesionless soils is believed to be the buildup of excess hydrostatic pressure due to the application of either a shock or cyclic loading. An evaluation of the effect of saturation on liquefaction behavior as reflected by the Skempton B-value for a loose sand has been presented by Chaney (1978). It was shown in this study that as B-value decreases (i.e., decreasing level of saturation) the number of cycles to liquefaction increases. As a consequence of the applied stresses, the structure of the cohesionless soil tends to become more compact with a resulting transfer of stress to the pore water and a reduction in stress on the soil grains. The soil grain structure, in response, rebounds to the extent required to keep the volume constant and this interplay of volume reduction and soil structure rebound determines the magnitude of increase in pore water pressure in the soil (Martin et al., 1975). This behavior is slightly modified by the effect of stress concentrations (Chaney, 1980). As the pore water pressure approaches a value equal to the applied confining pressure, the sand begins to undergo deformations. If the sand is loose, the pore pressure will increase suddenly to a value equal to the applied confining pressure, and the sand will rapidly begin to undergo large deformations. If the sand will undergo virtually unlimited deformations without mobilizing significant resistance to deformation, it can be said to be liquefied. In contrast, if the sand is dense, it may develop a residual pore water pressure after completion of one full cycle of loading which is equal to the confining pressure. On application of the next cycle of loading, or if the sand is subject to monotonic loading, the soil will tend to dilate. The corresponding pore water pressure will drop if the sand is undrained, and the soil will develop enough resistance to withstand the applied load. In order for this to occur, the soil will have to undergo some deformation to develop the resistance. If the cyclic loading continues, the amount of deformation required to produce a stable condition may *increase*. Ultimately, there will be a level of deformation at which the soil can withstand any load application. This type of behavior is termed cyclic mobility and is considerably less serious than liquefaction. In both cases there is a generation of excess pore water pressure, which must be dissipated. The dissipation of the pore pressure can affect the behavior of overlying sediments by the formation of sand boils. The movement of pore water due to excess pressure results in the erosion and movement of sediments from the liquefied zones to the soil surface. This migration of sediments leaves voids in the underlying soil stratum, which ultimately collapses due to the overburden weight and thus, causes the distortion of the surface. Due to the segregating action of wave attack on coastal materials and the resulting *seaward transport, most coastal deposits have* a relatively narrow particle size range as discussed by Fang and Chaney (1986). These relatively uniform deposits make them susceptible to quake liquefaction. In addition, the deposits void ratios normally exceed their critical void ratio (CVR) and, therefore, are in a potentially liquid state. These types of soils may be changed into actual macromeritic liquids (Winterkorn, 1953), throughout the whole granular system. A more detailed discussion of the liquefaction phenomena in the marine environment has been presented by Chaney and Fang (1991).

7.8.3.5.2 Variations in Liquefaction

There are two general classes of liquefaction phenomena depending on loading: (1) spontaneous liquefaction and (2) cyclic liquefaction. Spontaneous liquefaction is the result of the collapse of a metastable grain skeleton when subject to a mild shock. The mild shock can be generated by sudden differential settlement or because of a sudden rotational failure of a slope, or blasting effects.

In contrast, cyclic liquefaction is the result of the application of either one-way (compressional) or two-way (compressional/tension) loadings (Figure 7.25). Cyclic stresses can be the result of wave loading of seabed sediments, earthquake loading conditions, movement of jack-up platform legs to wave loading, or sustained vibrations from heavy equipment and rail traffic. A summary of these types of loadings is shown in Figure 7.25.

7.8.3.5.3 Wave Interaction With Seabed

The seafloor will be subjected to time-varying pressure in the nearshore region where the depth of water is not great and the waves can be considered stable as shown in Figure 7.26. The relationship between wave period (T), wavelength (L), and water depth (d) for a rigid bottom is presented in Figure 7.27 based on the equations in Figure 7.26. Pressure on the rigid seabed due to passage of water waves can be

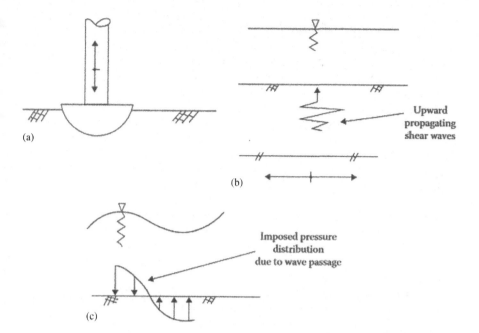

(a)

(b)

Upward
propagating
shear waves

(c)

Imposed pressure
distribution
due to wave passage

FIGURE 7.25 Typical types of cyclic loadings in the marine environment: (a) Jackup plat-form leg (one-way); (b) earthquake loading (two-way); and (c) ocean wave loading (two-way). **(Chaney and Fang, 1991)**

determined using Airy linear wave theory and assuming amplitude of waves is small relative to water depth and the seabed is rigid and impermeable. Based on Figure 7.28 the pressure is given by the following equation.

$$\Delta p = \Delta p' \sin 2\pi \left(\frac{x}{L} - \frac{t}{T} \right), \tag{7.52}$$

where

$$\Delta p' = \gamma_w \frac{H}{2} \left[\frac{1}{\cosh(2\pi d/L)} \right]. \tag{7.53}$$

The value of L and d is obtained from the Airy wave theory.

$$L = \left(\frac{gT^2}{2\pi} \right) \tanh \left(\frac{2\pi d}{L} \right). \tag{7.54}$$

The corresponding wave profile is given by the following:

$$y_s = \frac{H}{2} \sin 2\pi \left(\frac{x}{L} - \frac{t}{T} \right), \tag{7.55}$$

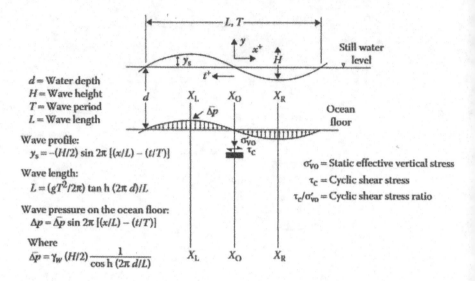

d = Water depth
H = Wave height
T = Wave period
L = Wave length

Wave profile:
$y_s = -(H/2) \sin 2\pi [(x/L) - (t/T)]$

Wave length:
$L = (gT^2/2\pi) \tan h (2\pi d)/L$

Wave pressure on the ocean floor:
$\Delta p = \bar{\Delta p} \sin 2\pi [(x/L) - (t/T)]$

Where
$\bar{\Delta p} = \gamma_w (H/2) \dfrac{1}{\cos h (2\pi d/L)}$

σ'_{vo} = Static effective vertical stress
τ_c = Cyclic shear stress
τ_c/σ'_{vo} = Cyclic shear stress ratio

FIGURE 7.26   Relationships between wave profile, wavelength, and wave pressure on rigid seafloor.

(Seed and Rahman, 1978)

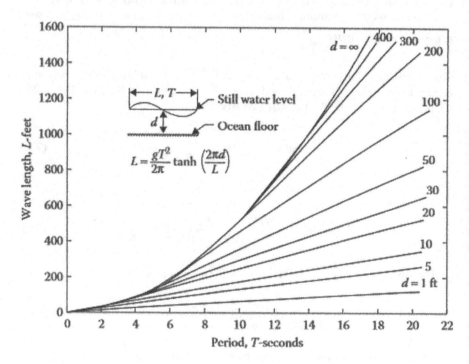

FIGURE 7.27   Relationship between wave period, wavelength, and water depth.

(Seed and Rahman, 1978)

where:

> H is the double amplitude of wave
> γ_w is the unit weight of water
> d is the depth of water
> L is the wavelength
> T is the wave period

A review of this figure shows that there is a unique relationship between water depth, period, and wavelength. In particular, Figure 7.27 shows that wave-induced bottom pressures become important in water depth less than approximately 152 m. Waves in water depths less than 152 m cause a progressive increase in pore water pressure. The rate and amount of pore pressure buildup will depend on several factors: (1) height, period, and lengths of different wave components; (2) cyclic loading characteristics of the seabed deposits; and (3) drainage and compressibility of the soil deposit.

The pressure Δp is given in Equation (7.52) and is plotted in Figure 7.28 as a dimensionless pressure amplitude against d/L. A review of Figure 7.28 shows that wave-induced bottom pressure are small when $d/L > 0.5$. The general ocean wave problem differs from the earthquake problem in four major aspects:

1. The storm waves have periods considerably longer than earthquake loadings.
2. The duration of ocean storms is significantly longer than that of earthquake cyclic loadings.
3. There is a high probability that a structure in the ocean will be subjected to a number of minor storms followed by relatively quiet periods before the occurrence of the maximum design loading conditions corresponding to a 100-year storm.

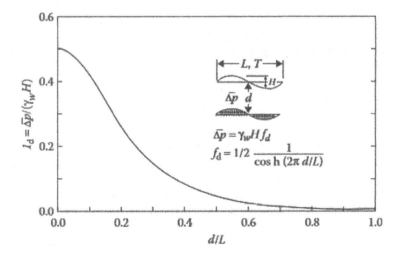

FIGURE 7.28 Wave-induced pressure on ocean floor.

(Seed and Rahman, 1978)

4. Wave loading is at the mudline, whereas shear waves induced by earthquakes propagate upward from a lower level in the ground. In contrast, the earthquake problem assumes implicitly the following:
   a. The maximum earthquake will be the first and perhaps the only significant seismic disturbance to affect the site during the lifetime of the structure.
   b. Because of the very short duration of an earthquake, there could be no drainage of any excess pore water pressure developed during the cyclic loading.

The behavior of the soil following liquefaction depends on the type of cyclic loading. For earthquakes that produce upward propagating waves, there is some argument that after liquefaction, shear stress waves cannot be transmitted and therefore, laboratory deformations following liquefaction in controlled stress cyclic loading tests may not be directly meaningful. However, the intensity of ocean wave-induced stresses would continue undiminished after liquefaction so that deformations after liquefaction would tend to be more significant for ocean wave than the earthquake problem.

The propagation of water waves over a permeable seabed exerts a time-varying pressure at the sediment–water interface. The time-varying pressure will cause cyclic variations in pore pressure and stresses within the bed. The effective stress varies in response to wave loading. Since soil strength is directly related to effective stress, any change in the effective stress state within the bed will affect bed strength and stability. Many coastal structures such as pipelines, platforms, anchors, and breakwaters that interact with the seabed will be affected by both cyclic effective stresses and the erosion potential of the bottom sediments. There have been a number of analytical studies that have examined the hydraulics of waves interacting with the seabed under a variety of conditions. Some of these conditions are the following:

- Saturated beds with isotropic permeability.
- Saturated anisotropic permeability.
- Stratified permeability.
- Unsaturated or compressible seabed.
- Effective stress state within bed.

The pressures calculated using the above approaches generally use the linear (Airy) theory and expect the pore pressure response to be in phase with the surface waves. This is in agreement with what has been measured for sandy bottoms. In contrast, measurements made by Hirst and Richards (1977) showed that bottom pressures can be much larger than predicted by linear theory for soft clayey bottoms. In the following the cases of a rigid seabed (i.e., sandy bottom) will be considered first followed by the case of a deformable seabed (i.e., soft clayey bottom).

7.8.3.5.4 Rigid Seabed

The wave-induced pressure on a rigid ocean floor for a sinusoidal wave has been previously presented in Figure 7.28 (Seed and Rahman, 1978). The wave-induced

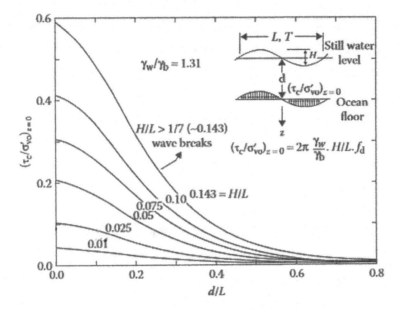

FIGURE 7.29 Relationship between wave characteristics and wave-induced shear stress.
(Seed and Rahman, 1978)

pressures were shown to decrease with depth as given by Equation (7.53). These wind–wave-induced bottom pressures become important in water depths less than 152 m. A review of Figure 7.28 shows that the water pressure drops off exponentially.

Wave-induced shear stresses can be evaluated using the theory of elasticity. Simple charts for evaluating the shear stress ratio developed at any depth for waves having different characteristics are presented in Figure 7.29. The development of high pore pressures caused by the action of waves on an environment involving sand deposits can lead to instability. This instability is an important concern for many engineering installations, such as pipelines and anchors.

Seed and Rahman (1978) developed a procedure for evaluating the magnitude and distribution of wave-induced pore pressures in ocean floor deposits. The generation of excess pore pressure in terms of the number of cycles NL required to cause initial liquefaction under the given stress conditions has been presented by Seed and Booker (1977) in Equation (7.56).

$$\frac{u_g}{\sigma'_{vo}} = \left(\frac{2}{\pi}\right) \arcsin\left(x^{1/2\theta}\right), \tag{7.56}$$

where:

σ'_{vo} is the initial vertical effective stress
θ is the empirical constant, average value 0.7
$x = N/N_L$

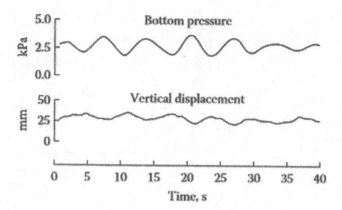

FIGURE 7.30 Simultaneous records of the wave-induced bottom pressure and the vertical displacement.

(Suhayda, 1977)

7.8.3.5.5 Deformable Seabed

Part of the SEASWAB project was to acquire direct measurements of wave action and seafloor response of marine clays under wave action. Suhayda (1977) reported on field measurements of bottom oscillations and wave characteristics in a study of the interaction of fine-grained sediments and surface waves (Figure 7.30). This was accomplished by using a wave staff, pressure transducers, and accelerometer in East Bay, Louisiana. This area has a fine-grained clay bottom. The bottom sediments appeared to be undergoing an elastic response to surface wave crest. This was evidenced by the bottom being depressed by a surface wave crest. Under the range of bottom pressures measured, Suhayda (1977) reported that bottom displacement varied linearly with bottom pressures. The measured bottom pressures were found to be up to 35% larger than that predicted by using linear (Airy) wave theory. Therefore, the energy lost from the surface wave to the bottom is significant and larger than the energy lost to bottom friction. This relationship is shown as the ratio of bottom pressure for a moving bottom to the bottom pressure without any motion (Airy wave) versus relative depth as a function of the $B = b/a$ in Figure 7.31. The value a is the maximum amplitude of the surface wave while b is the maximum amplitude of the bottom sediment motion. A review of this figure shows that for an out of phase wave an elastic-like response is generated resulting in a greater bottom pressure than if the waves are in phase. In summary, for reasonable wave heights, a flexible bottom dissipates energy at a rate at least an order of magnitude greater than a rigid impermeable bottom (Suhayda, 1977).

A relatively greater amount of wave energy is lost on a muddy coast at intermediate water depths than is dissipated on a sandy coast (Suhayda, 1977).

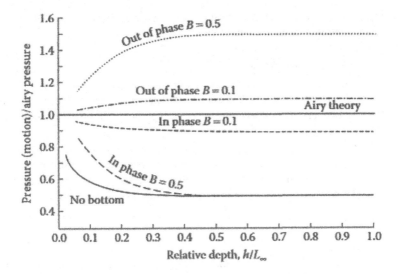

FIGURE 7.31 Ratio of the predicted bottom pressure over a moving bottom to the bottom pressure predicted from linear wave theory, as a function of relative depth. B is the ratio of b/a.

(Suhayda, 1977)

7.8.3.5.6 Angle of Repose

A cone-shaped mound occurs when a dry sand-sized sediment is poured onto a flat surface. The maximum slope of the cone at equilibrium is called the angle of repose. The angle of repose concept is of interest for a number of reasons. These include the stability of rubble-mound breakwaters and the modeling of sediment transport. The angle of repose is a function of grain shape and increases with increasing grain irregularity. The variation of the angle of repose in relationship to various sediment characteristics has been studied by a number of investigators: Allen (1970), Cornforth (1973), Lane (1955), Simons and Albertson (1960), Simons and Senturk (1977), and Statham (1974).

Bagnold used the relationship between the shear forces, the normal forces, and the angle of repose to develop his energetics-based sediment transport model. He reasoned that the amount of sediment by weight at the top of a bed that can be supported in an elevated state and then be transported is related to the applied shear stress of the overlying (moving) fluid by the tangent of the angle of repose. Bagnold's concept forms the basis of many of the sediment transport models most frequently used by coastal engineers. For further discussion, see Bagnold (1963, 1966), Bailard (1981), and Bailard and Inman (1979).

7.8.4 BULK PROPERTIES OF DIFFERENT SEDIMENTS

7.8.4.1 Clays, Silts, and Muds

Engineers encounter clay in cliffs, foundations or as a material to be dredged. The coastal plains, bays, and lagoons are often underlain by clay. Older clays are often consolidated and can stand with near-vertical slopes when eroded. As an example, the deepest parts of tidal inlets may have steep sides cut into stiff clay. Many eroding coastal flats contain a large amount of clay.

7.8.4.1.1 Silt-Sized Particle

Silt-sized particles have sizes intermediate between sand and clay. The majority of silt is produced by either the gradual chemical weathering of rocks, or as rock flour ground out by glaciers. Sediments consisting primarily of silt are common in deltas, and estuaries.

Silt is usually absent on beaches where waves drive the dominant processes. Silt is easily removed from any shore where wave action is moderate or severe because it remains in suspension far longer than sand grains. Silt has very little cohesion when dry and will easily fall apart. In contrast, clay when dry will have the consistency of a brick.

7.8.4.1.2 Muds

Muds are watery mixtures of clay and silt often with minor amounts of sand and organic material. Muds behave similar to a viscous fluid than as a cohesive solid. Muds often accumulate in dredged channels where the upper and lower boundaries of the mud layer can be difficult to determine.

7.8.4.1.3 Organically Bound Sediment

Marsh grasses bind sands, silts, and clays with their root systems to form an organically bound sediment. These grasses typically grow in back bays and other tidal wetlands. They are sometimes called peat (typically, organic silt in the engineer's soil classification). These sediments are very compressible. Field evidence indicates that when these type of sediments are overlain by a barrier island the compression of organic matter by the weight of the sand results in subsidence.

Organically bound sediment along the shoreline can be exposed to ocean wave erosion. Erosion of the organically bound sediment by ocean waves produces pillow- or cobble-shaped fragments. These fragments of organically bound sediment are often found on barrier island coasts after storms.

The bulk density values of organically bound sediment are within the expected range as shown in Table 7.11.

7.8.4.1.4 Sand and Gravel

Ocean beaches typically consist of sand and some gravel. *The temperate zone ocean beach is typically made of quartz sand.* The median diameter D50 of the sand is in the range $0.15 < D50 < 0.40$ mm. In colder latitudes the composition of the beach material becomes more varied and size tends to increase. This is caused by geologically recent glacial action that has affected sediment supply and the wave climate is more severe. In these northern latitudes, sand grains include silicate and heavy minerals along with the quartz. In addition, coarse pebbles (gravel or "shingle") may locally dominate the sediment.

7.8.4.1.5 Cobbles, Boulders, and Bedrock

Tectonically active areas and shorelines often have abundant cobbles, boulders, and bedrock. These types of shores result from abundant supplies of rock produced by a combination of glaciers, stream erosion of mountainous terrain, and enhanced wave erosion at the end of long fetches.

7.8.4.2 Bays and Estuary Beaches

Bays and estuary beaches at the same latitudes often differ from ocean beaches. The beach material along interior shores of waterways tends to be coarser and more limited in volume than along ocean shores. This is probably because of the smaller wave action on interior waters has not eroded enough material to produce extensive sand beaches.

7.9 FALL VELOCITY

7.9.1 INTRODUCTION

A particle falling through either water or air accelerates until it reaches its terminal velocity. This is the velocity that a particle reaches when the drag force on the particle just equals the downward gravitational force.

A particle's fall velocity is a function of its size, shape, and density; as well as the fluid density, viscosity, temperature, and a number of other parameters.

The terminal or settling velocity of natural sediment particles has been extensively studied since the early work of Stokes in 1851. The vast variation of particle geometries of natural sediment particles introduces significant difficulties in an attempt to establish a similar relationship as already done previously for spherical particles.

Three equations describing the relationship between the fall velocity and the Archimedes buoyancy index, with each equation for a specific flow regime have been developed by Hallermeier (1981). These equations are based on a collection of previous experimental results for natural particles such as sand particles. Using a similar approach, Ahrens (2000) developed a best fit solution consisting of a single equation.

7.9.2 GENERAL EQUATION

The balance between the drag force and the gravitational force for a single sphere falling in a still fluid, is the following:

$$C_D \frac{\pi D^2}{4} \frac{\rho W_f^2}{2} = \frac{\pi D^3}{6}(\rho_s - \rho)g. \tag{7.57}$$

Rearranging and solving for the fall velocity:

$$W_f = \left(\frac{4}{3} \frac{gD}{C_D} \left[\frac{\rho_s}{\rho} - 1 \right] \right)^{1/2}, \tag{7.58}$$

where:

W_f = fall velocity $(gD)^{1/2}$
C_D = dimensionless drag coefficient
D = grain diameter
ρ = density of water
ρ_s = density of the sediment

The problem now is determining the appropriate drag coefficient. The drag coefficient (C_D) has been shown to vary as a function of the Reynolds number $(Re = W_f D/\upsilon$, where υ is the kinematic viscosity) for spherical particles. The drag coefficient versus Re is presented in Figure 7.32 based on data from Rouse and others. Re is dimensionless, but W_f, D, and υ must have common units of length and time.

The plot in Figure 7.32 can be divided into three regions depending on Re. In the linear region, Re is shown to be less than approximately 0.5. The drag coefficient in this region is shown to decrease linearly with Reynolds number. This linear region represents small, light grains falling at slow velocities.

The drag on these small grains is dominated by viscous forces, rather than inertia forces. The corresponding fluid flow past the particle is entirely laminar. In this region, Stokes found the analytical solution for C_D was the following:

$$C_D = \frac{24}{Re} = \frac{24\upsilon}{W_f D}. \tag{7.59}$$

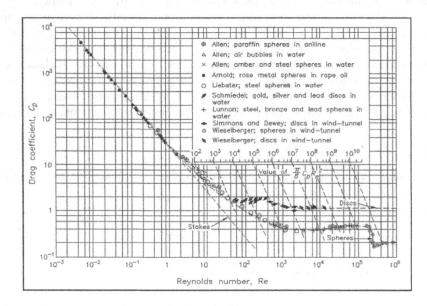

FIGURE 7.32   Drag coefficient CD as a function Reynolds number.

(U.S. Army Corps of Engineers, *Coastal Engineering Manual*, 2008. Courtesy of the U.S. Army)

This line is labeled "Stokes" in Figure 7.32. Substituting Equation (7.59) into Equation (7.58) gives the fall velocity of particles in this region:

$$W_f = \frac{gD^2}{18v}\left(\frac{\rho_s}{\rho} - 1\right). \tag{7.60}$$

In this region the fall velocity increases as the square of the grain diameter, and is a function of the kinematic viscosity.

The intermediate Reynolds number range is from $Re > 400$ to $Re < 200{,}000$. The drag coefficient (C_D) in this range has an approximate value ranging from 0.4 to 0.6. The size of the particles in this Reynolds number range are larger and denser, and therefore the fall velocity increases. The reason for this change in C_D is that inertial drag forces have become predominant over the viscous forces, and the wake behind the particle has become turbulent. For this region the approximation $C_D \sim 0.5$ is used in Equation (7.58) to obtain the following fall velocity:

$$W_f = 1.6\left(gD\left[\frac{\rho_s}{\rho} - 1\right]\right)^{1/2}. \tag{7.61}$$

A review of Equation (7.61) shows that the fall velocity varies as the square root of the grain diameter and is independent of the kinematic viscosity.

At approximately $Re = 200{,}000$ the drag coefficient decreases abruptly. This is the region of large particles that have a high fall velocity. The wake of the falling particle is turbulent. In addition, the flow in the boundary layer around the particle is turbulent as well. Similarly, in the region $Re > 200{,}000$, the approximation $C_D \sim 0.2$ is used in Equation (7.58) to obtain the following fall velocity:

$$W_f = 2.6\left(gD\left[\frac{\rho_s}{\rho} - 1\right]\right)^{1/2}. \tag{7.62}$$

There is a large transition region between the first two regimes (between $0.5 < Re < 400$). These Reynolds numbers correspond to quartz grain sizes between about 0.08 mm and 1.9 mm falling in water. This closely corresponds to all sand particles, as seen in Table 7.5. Thus, for very small particles (silts and clays), the fall velocity is proportional to $D^{1/2}$ and can be calculated from Equation (7.60). For gravel size particles the fall velocity is proportional to $D^{1/2}$ and can be calculated from Equation (7.61). In contrast, for sand, no simple formula is available. The fall velocity for sand is in a transition region between a D^2 dependence and a $D^{1/2}$ dependence. In this size range, a fall velocity value can be obtained from plots such as Figure 7.33. This figure shows the fall velocity as a function of grain diameter and water temperature for quartz spheres falling in both water and air. The vertical and horizontal axes are grain diameter and fall velocity. The short straight lines crossing the curves obliquely are various values of Re. The transition between the second and third regime corresponds to approximately 90-mm quartz spheres (cobbles, as shown on Table 7.5) falling in water.

For grain sizes outside the range shown in Figure 7.33, Equation (7.58) can be used with a value of C_D obtained from Figure 7.32. This is an iterative procedure involving repeated calculations of fall velocity and Reynolds number. Instead, Equation (7.58) can be rearranged to yield:

$$\frac{\pi}{8} C_D Re^2 = \frac{\pi D^3 \left(\frac{\rho_s}{\rho} - 1 \right) g}{6v^2}. \tag{7.63}$$

This quantity, $\pi/8 \, C_D \, Re^2$, can be used in Figure 7.32 to obtain a value of C_D or Re, these can then be used to calculate the fall velocity.

7.9.3 EFFECT OF DENSITY

The fall velocity calculated using Equation (7.58) is dependent on the following factor $((\rho_s/\rho) - 1)^{1/2}$. For quartz sand grains in fresh water, this factor is about 1.28. In contrast, in ocean water, this factor reduces to about 1.25 because of the slight increase in density of salt water.

7.9.4 EFFECT OF TEMPERATURE

Increasing temperature has the effect of decreasing the density of water. This result is small compared to temperatures effect on the coefficient of viscosity. Changes in the fluid viscosity affect the fall velocity for small particles, but not for large. (Equation (7.59) contains a viscosity term while Equations (7.60) and (7.61) do not.) Figure 7.33 has separate lines labeled with temperatures of 0°C, 10°C, 20°C, 30°C, and 40°C, and these lines are reversed for the two fluids (i.e., water and air). In summary, a grain will fall faster in warmer water, but slower in warmer air. This is a result of the viscosity variation with temperature in the two fluids.

7.9.5 EFFECT OF PARTICLE SHAPE

Grain shape affects the fall velocity of large particles (those larger than Re about 10 or D about 0.125 mm in water). In contrast, grain shape has a negligible effect on small particles. The less spherical the particle, the slower the fall velocity for large grains of a given nominal diameter. In Figure 7.32 it is seen that disks have higher C_D values, and thus lower fall velocities than spheres at large Reynolds numbers as shown in Figure 7.32.

7.9.6 OTHER FACTORS

The fall velocity is affected by several other factors.

- A tight clump of grains in an otherwise clear fluid will fall faster than a single grain. This is because the adjacent fluid is partially entrained and thus the drag on each particle decreases.

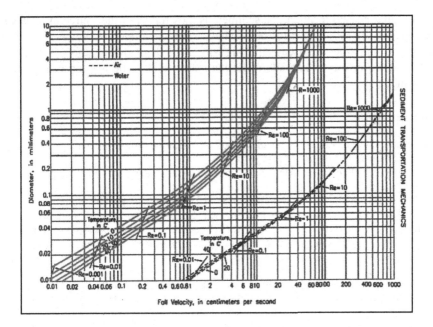

FIGURE 7.33   Fall velocity of quartz spheres in air and water.

(Vanoni, 1975)

- If the grains are uniformly distributed in the fluid, each will fall slower. This is because, as each grain falls, replacement fluid must flow up and this flow impedes the other grains.
- An adjacent wall of the testing apparatus will decrease the fall velocity.

These effects need to be considered in the design and calibration of a settling tube used to measure the relationship between grain diameter and fall velocity.

7.10   SUMMARY

The materials that make up the coast are fundamental in determining the nature of the coast. One of the most fundamental factors is whether the geologic material comprising the coast is dominated either by bedrock or by sediment. Sediments can be further divided into cohesionless material (i.e., sand and gravel) or cohesive (i.e., clays). The movement of these materials along the coastline occurs by a wide variety of processes. This variety of different processes occurs over a wide range in intensity. Together these different processes are important factors in determining the morphology and scale of the various coastal environments.

Low-energy processes (i.e., small waves) tend to accumulate sediments that are fine grained and/or those that are not well sorted. More energetic conditions (i.e., large waves, or earthquake loading) result in sand and gravel that tends to be well sorted. Regardless of the nature or intensity of the processes operating along a particular coastal location, the sediment is the product of the source materials in that vicinity. In this chapter the physical and chemical properties of materials found along the shoreline are discussed along with methods of classification. In addition, the behavior of these materials under both static and cyclic loading conditions is presented.

REFERENCES

Ahrens, B. (2000). New developments in soil classification. World Reference Base for Soil Resources. *Geoderma*, 96, 345–357.

Allen, J. R. L. (1970). The avalanching of granular solids on dune and similar slopes. *Journal of Geology*, 78, 326–351.

American Society for Testing Materials (ASTM). (1994). Volume 04.08, Soil and Rock (1): D420 – D4914, Philadelphia, PA: American Society for Testing and Materials.

Aylmore, L. A. G., and Quirk, J. P. (1959). Swelling of clay systems. *Nature*, 183, 1752–1753.

Bagnold, R. A. (1963). Beach and nearshore processes; Part I: Mechanics of marine sedimentation. In Hill, M. N. (Ed.), *The Sea: Ideas and Observations*. New York: Interscience, Vol 3, pp. 507–528.

Bagnold, R. A. (1966). An Approach to the Sediment Transport Problem from General Physics, U.S. Geological Survey Professional Paper 422-I, U. S. Department of the Interior.

Bailard, J. A. (1981). An energetics total load sediment transport model for a plane sloping beach. *Journal of Geophysical Research*, 86(C11), 10938–10954.

Bailard, J. A., and Inman, D. L. (1979). A reexamination of Bagnold's granular fluid model and bed load transport. *Journal of Geophysical Research*, 84(C12), 7827–7833.

Bennett, R. H. (1976). Clay fabric and geoteclinical properties of selected submarine sediment cores from the Mississippi Delta. Ph.D. dissertation, Texas A&M University, 269 p.

Bennett, R. H., and Hulbert, M.H. (1986). Concepts of clay fabric. In *Clay Microstructure*. Geological Sciences Series. Dordrecht: Springer, 47–103.

Bennett, R. H., Fischer, K. M., Lavoie, D. L., Bryant, W. R., and Rezak, R. (1989). Porometry and fabric of marine clay and carbonate sediments: Determinants of permeability. *Marine Geology*, 89(1–2), 127–152.

Berger, W. H., and von Rad, U. (1972). Cretaceous and Cenozoic sediments from the Atlantic Ocean. In Hayes, D. E., Pimm, A.C., et al. (EDS.), *Initial Reports of the Deep Sea Drilling Project*, Volume 14. Washington, D.C.: U.S. Government Printing Office), pp. 787–954.

Bishop, A. W., and Eldin, A. K. G. (1950). Undrained triaxial tests on saturated sands and their significance in the general theory of shear strength. *Geotechnique*, 2, 129–150.

Bjerrum, L., and Simons, N. E. (1960). Comparison of shear strength characteristics of normally consolidated clays. *Norwegian Geotechnical Institute*, 35, 13–22.

Blatt, H., Middleton, G., and Murray, R. (1980). *Origin of Sedimentary Rocks*. 2nd ed. Englewood Cliffs, NJ: Prentice-Hall.

Busch, W. H., and Keller, G. H. (1981). The physical properties of Peru-Chile continental margin sediments—The influence of coastal upwelling on sediment properties. *Journal of Sedimentary Petrology*, 51, 705–719.

Casagrande, A. (1932). Research on the Atterberg limits of soil. *Public Roads*, 13, 121–136.

Casagrande, A. (1948). Classification and identification of soils. *Transactions of the American Society of Civil Engineers*, 113, 901–930.

Chamberlain, T. K. (1960). Mechanics of mass sediment transport in Scripps Submarine Canyon. Ph.D. dissertation, Scripps Institution of Oceanography, University of California, San Diego.

Chaney, R. C. (1978), Saturation effects on the cyclic strengths of sand. Proceedings of the Specialty Conference on Earthquake Engineering and Soil Dynamics, ASCE, Pasadena, CA, 1, pp. 342–358.

Chaney, R. C. (1980). Seismically induced deformations in earthdams. *7th World Conference on Earthquake Engineering*, 31, 482–486.

Chaney, R. C., and Almagor, G. (2016). *Seafloor Processes and Geotechnology*. Boca Raton, FL: CRC Press, p. 558.

Chaney, R. C., and Fang, H. Y. (1986). Static and dynamic properties of marine sediments: A state of the art. In Chaney, R. C., and Fang, H. Y. (Eds.), *Marine Geotechnology and Nearshore/Offshore Structures*. Philadelphia, PA: ASTM Press, 74–111.

Chaney, R. C., and Fang, H. Y. (1991). Liquefaction in the coastal environment: An analysis of case histories. *Marine Geotechnology*, 10(3–4), 343–370.

Chaney, R. C., Slonim, S. M., and Slonim, S. S. (1982). Determination of calcium carbonate content in soils. In Demars, K. R., and Chaney, R. C. (Eds.), Geotechnical Properties, Behavior, and Performance of Calcareous Soils, ASTM STP 777, American Society for Testing and Materials, pp. 3–15.

Chassefiere, B., and Monaco, A. (1983). On the use of Atterberg limits on marine soils. *Marine Georesources and Geotechnology*, 5(2), 153–179.

Conley, D. C., and Inman, D. L. (1992). Field observations of the fluid-granular boundary layer under near-breaking waves. *Journal of Geophysical Research*, 97(C6), 9631–9643.

Cornforth, D. H. (1973). Prediction of Drained strength of sands from relative density measurements. American Society for Testing and Materials, Spec. Tech. Pub. 523, pp. 281–303.

Daly, R. A., Manger, G. E., and Clark, S. P. Jr. (1966). Density of rocks. In Clark, S. P., Jr. (Ed.), *Handbook of Physical Constants*. Revised ed., pp. 19–26.

Darcy, A. (1856). *Les Fontaines Publiques de la Ville de Dijon*. Paris, France: Dalmont.

DeAlba, P., Seed, H. B., and Chan, C. K. (1976). Sand liquefaction in largescale simple shear tests. *Journal of the Geotechnical Engineering Division*, 102(9), 909–927.

Dill, R. F. (1964). Sedimentation and erosion in Scripps submarine canyon head. In Miller, R. I. (Ed.), *Papers in Marine Geology, Shepard Commemorative Volume*. New York: MacMillan & Co., pp. 23–41.

Fang, H. Y., and Chaney, R. C. (1986). Geo-environmental and climatological conditions related to coastal structural design along the China coastline. *Proceedings of the Symposium on Marine Geotechnology and Nearshore/Offshore* Structures, Shanghai, China, ASTM STP 923, pp. 149–160.

Farrar, D. M., and Coleman, J. D. (1967). The correlation of surface area with other properties of nineteen British clay soils. *Journal of Soil Science*, 18(1), 118–124.

Finn, W. D. L., and Bhatia, S. K. (1981). Prediction of seismic pore-water pressures. In *Proceedings, of the 10th International Conference on Soil Mechanics and Foundation Engineering 3*, Rotterdam, the Netherlands, pp. 201–206.

Folk, R. L. (1951). Stages of textural maturity in sedimentary rocks. *Journal of Sedimentary Petrology*, 21, 127–130.

Folk, R. L. (1965, 1974). *Petrology of Sedimentary Rocks*. Austin, TX: Hemphill Publishing Company.

Folk, R. L. (1966). A review of grain-size parameters. *Sedimentology*, 6, 73–93.

Folk, R. L. (1974). *Petrology of Sedimentary Rocks*. Austin, TX: Hemphill Publishing Company.

Folk, R. L., and Ward, W. C. (1957). Brazos River Bar. A study in the significances of grain size parameters. *Journal of Sedimentary Petrology*, 27, 3–26.

Galvin, C. (1979). Relation Between Immersed Weight and Volume Rates of Longshore Transport, Technical Paper 79-1, Coastal Engineering Research Center, U.S. Army Engineer Waterways Experiment Station, Vicksburg, MS.

Galvin, C., and Alexander, D. F. 1981. Armor Unit Abrasion and Dolos Breakage by Wave-Induced Stress Concentrations, American Society of Civil Engineers, Preprint 81–172.

Gibson, R. E. (1958). The progress of consolidation in a clay layer increasing in thickness with time. *Geotechnique*, 8, 171–182.

Griffiths, J. C. (1967). *Scientific Method in Analysis of Sediments*. New York: McGraw-Hill.

Grim, R. E. (1968). *Clay Mineralogy*, 2nd ed. New York: McGraw-Hill, p. 596.

Grimshaw, R. W. (1971). *The Chemistry and Physics of Clays*, 4th ed. London: Ernest Benn Publishers.

Hallermeier, R. J. (1981). Terminal settling velocity of commonly occurring sand grains. *Sedimentology*, 28(6), 859–865.

Handin, J. (1966). Strength and ductility. In Clark, S. P., Jr. (Ed.), *Handbook of Physical Constants*. Geological Society of America, Memoir 97, pp. 223–289.

Hirst T. J., and Richards A. F. (1977). *In situ* Pore Pressure Measurements in Mississippi Delta Front Sediments, *Marine Geotechnology* 2, 191–204.

Hobson, R. D. (1977). *Review of Design Elements for Beach-Fill Evaluation*, TP 77-6. Vicksburg, MS: Coastal Engineering Research Center, U.S. Army Engineer Waterways Experiment Station.

Hunt, J. M. (1979). *Petroleum Geochemistry and Geology*. San Francisco, CA: W.H. Freeman.

Inman, D. L. (1952). Measures for Describing the Size Distribution, pp. 125–145.

Inman, D. L. (1957). Wave Generated Ripples in Nearshore Sands, Technical Memorandum 100, Beach Erosion Board, U.S. Army Corps of Engineers.

Jakobson, B. (1953). Origin of cohesion of clay. *Proceedings of the 3rd International Conference on Soil Mechanics and Foundation Engineering*, Zurich, Switzerland.

Johnson, G. R., and Olhoeft, G. R. (1984). Density of rocks and minerals. In Carmichael, R. S. (Ed.), *Handbook of Physical Properties of Rocks, Vol III*. CRC Press, Inc., pp. 1–28.

Keller, G. H. (1968). Shear strength and other physical properties of sediments from some ocean basins. *Proceedings of the Conference on Civil Engineering in the Oceans*, ASCE Press, San Francisco, CA, pp. 319–417.

Kenney, T. C., and Folkes, D. J. (1979). Mechanical properties of soft soils. State-of-the-Art report to session 2, *32nd Canadian Geotechnical Conference*, Quebec, Canada.

Krinsley, D. H., and Doornkamp, J. C. (1973). *Atlas of Quartz Sand Surface Textures*. Cambridge: Cambridge University Press.

Kranck, K. (1980). Experiments on the significance of flocculation in the settling of fine-grained sediment in still water. *Canadian Journal of Earth Sciences*, 17, 1517–1526.

Krumbein, W. C. (1936). Application of logarithmic moments to size frequency distribution of sediments. *Journal of Sedimentary Petrology*, 6(1), 35–47.

Krumbein, W. C. (1941). Measurement and geological significance of shape and roundness of sedimentary particles. *Journal of Sedimentary Petrology*, 11(2), 64–72.

Krumbein, W. C., and Monk, G. D. (1942). Permeability as a Function of the Size Parameters of Unconsolidated Sand, American Institute Mining and Metallurgy Engineering, Technical Publication No. 1492, Petroleum Technology, pp. 1–11.

Krumbein, W. C., and Sloss, L. L. (1963). Chapter 4, Properties of sedimentary rocks. *Stratigraphy and Sedimentation*. W. H. Freeman & Company, pp. 93–149.

Ladd, C. C. (1991). Stability evaluation during staged construction. *Journal of Geotechnical Engineering*, 117(4), 540.

Ladd, C. C., and Foote, R. 1974. New design procedure for stability of soft clays. *Journal of Geotechnical Engineering Division*, 100(7), 763–786.

Lambe, T. W., and Whitman, R. V. (1969). *Soil Mechanics*. New York: John Wiley.

Lane, E. W. (1955). Design of stable channels. *Transactions, American Society of Civil Engineers*, 120, 1234–1279.

Lisitzin, A. P. (1972). Sedimentation in the world ocean. In *Society of Economic Paleontologists and Mineralologists*, Special Publication 17, Tulsa, OK.

Manger, G. E. (1966). Porosity and bulk density, dry and saturated, of sedimentary rocks. In Clark, S. P., Jr. (Ed.), *Handbook of Physical Constants*, Geological Society of America Memoir 97, Table 4-4, pp. 23–25.

Margolis, S. V. (1969). Electron microscopy of chemical solution and mechanical abrasion features on quartz sand grains. *Sedimentary Geology*, 2, 243–256.

Martin, G. R., Finn, W. D. L., and Seed, H. B. (1975). Fundamentals of liquefaction under cyclic loading, *Journal of the Geotechnical Engineering Division, ASCE, 101*, No. GT-5, 423–438.

Mason, M. A. (1942). Abrasion of Beach Sands, TM-2, U.S. Army Corps of Engineers, Beach Erosion Board, Washington, DC.

McCammon, R. B. (1962). Efficiencies of percentile measures for describing the mean size and sorting of sedimentary particles. *Journal of Geology*, 70, 453–465.

McLean, R. F., and Kirk, R. M. (1969). Relationship between grain size, size-sorting, and foreshore slope on mixed Sandy-shingle beaches. *New Zealand Journal of Geology and Geophysics*, 12, 128–155.

Michaels, A. S., and Lin, C. S. (1954). Permeability of Kadinite. *Industrial and Engineering Chemistry*, 46, 1239–1246.

Mitchell, J. K. (1976). *Fundamentals of soil behavior.* New York: Wiley, p. 422.

Moon, C. F. (1972). The microstructure of clay sediments. *Earth Science Reviews*, 8(3), 303–323.

Moum, J., and Rosenquvist, I. T. (1961). The mechanical properties of montmorillonitic and illitic clays related to the electrolytes of the pore water. *Proceedings of the 5th International Conference on Soil Mechanics and Foundation Engineering*, 1, 263–267.

Mysels, K. J. (1959). *Introduction to Colloid Chemistry.* New York: Interscience Publishers.

Nacci, V. A., Kelly, W. E., Wang, M. C., and Demars, K. R. (1974). Strength and stress-strain characteristics of cemented deep-sea sediments. In Inderbitzen, A. L. (Ed.), *Deep-Sea-Sediments: Physical and Mechanical Properties.* New York: Plenum Press.

Noorany, I. (1989). Classification of marine sediments. *Journal of Geotechnical and Geoenvironmental Engineering*, 115, 23–37.

O'Brien, N. R. 1970. The fabric of shale—An electron-microscope study. *Sedimentology*, 15, 229–246.

Odell, R. T., Thornburn, T. H., and McKenzie, I. (1960). Relationships of Atterberg limit to some other properties of Illinois soils. *Proceedings of the Soil Science Society of America*, 24(5), 297–300.

Olsen, H. W. (1965). Deviations from Darcy's Law in saturated clays. *Soil Science Society of America, Proceedings*, 29, 135–140.

Otto, G. H. (1939). A modified logarithmic probability graph for the interpretation of mechanical analyses of sediments. *Journal of Sedimentary Petrology*, 9, 62–76.

Packwood, A. R. (1983). The influence of beach porosity on wave uprush and backwash. *Coastal Engineering*, 7, 29–40.

Packwood, A. R., and Peregrine, D. H. (1980). The propagation of solitary waves and bores over a porous bed. *Coastal Engineering*, 3, 221–242.

Pettijohn, F. J. (1957). *Sedimentary Rocks.* New York: Harper and Brothers, p. 117.

Powers, M. C. (1953). A new roundness scale for sedimentary particles. *Journal of Sedimentary Petrology*, 23, 117–119.

Pusch, R. (1973a). Influence of organic matter on the geotechnical properties of clays. National Swedish Building Research, Document Dll, Stockholm.

Pusch, R. (1973b). Influence of salinity and organic matter on the formation of clay microstructure. *Proceedings of the International Symposium on Soil Structure*, Gothenburg, Sweden. Swedish Geotechnical Society, Stockholm, Sweden, pp. 165–175.

Ramsey, M. D., and Galvin, C. (1977). Size Analyses of Sand Samples from Southern New Jersey Beaches, Miscellaneous Paper 77-3, Coastal Engineering Research Center, U.S. Army Engineer Waterways Experiment Station, Vicksburg, MS.

Reid, R. O., and Kajiura, K. (1957). On the damping of gravity waves over a permeable sea bed. *Transactions of the American Geophysical Union*, 38(5), (October), pp. 662–666.

Reimers, C. E. (1982). Organic matter in anoxic sediments off central Peru: Relations of porosity, microbial decomposition and deformation properties. *Marine Geology*, 46, 175–197.

Ritter, D. F. (1986). *Process Geomorphology*. 2nd ed. Dubuque, IA: Wm. C. Brown.

Savage, R. P. (1958). Wave run-up on roughened and permeable slopes. *Journal of the Waterways and Harbors Division, American Society of Civil Engineers*, 84(WW3), Paper 1640.

Seed, H. B., and Booker, J. R. (1977). Stabilization of potentially liquefiable and deposits using gravel drains. *Journal of the Geotechnical Engineering Division, ASCE*, 103/7, 757–768.

Seed, H.B., Noorany, I., and Smith, I. M. (1964). Effects of sampling and disturbance on the strength of soft clays. Report No. TE-64-1, Institute of Engineering Research, University of California, Berkeley, CA, p. 83.

Seed, H. B., and Rahman, M. S. (1978). Wave-induced pore pressure in relation to ocean floor stability of cohesionless soils. *Marine Geotechnology*, 3(2), 123–150.

Shepard, F. P. (1954). Nomenclature based on sand-silt-clay ratios. *Journal of Sedimentary Petrology*, 24, 151–158.

Shepard, F. P., and Young, R. (1961). Distinguishing between beach and dune sands. *Journal Sed. Pet*, 31(2), 196–214.

Shorten, G. G. (1993). Stratigraphy, sedimentology and engineering aspects of Holocene organo-calcareous silts, Swan Harbour, Fiji. *Marine Geology*, 110, 275–302.

Shorten, G. G. (1995). Quasi-overconsolidation and creep phenomena in shallow marine and estuarine organo-calcareous silts, Fiji. *Canadian Geotechnical Journal*, 32(1), 89.

Silva, A. J., Hetherman, J. R., and Calnan, D. I. (1981). Low-gradient permeability testing of fine-grained marine sediments. *American Society for Testing and Materials, Special Technical Publication*, 121–136.

Simons, D. B., and Albertson, M. L. (1960). Uniform water conveyance channels in alluvial material. *Journal Hydraulic Division, American Society of Civil Engineers*, 86, 33–99.

Simons, D. B., and Senturk, F. (1977). Sediment Transport Technology, Water Res. Pub., Fort Collins, CO.

Skempton, A. W. (1954). The pore-pressure coefficient A and B. *Geotechnique*, 4, 143–147.

Skempton, A. W. (1964). Long term stability of clay slope. *Geotechnique*, 14(2), 77–101.

Soderblom, R. (1969). Salt in Swedish clays and its importance for quick clay formation. *Swedish Geotechnical Proceedings*, 22, 63.

Sowers, G. B., and Sowers, G. F. (1961). *Introductory Soil Mechanics and Foundations*, 2nd ed. New York: Macmillan, p. 386.

Statham, I. (1974). *Ground Water Hydrology*. New York: John Wiley & Sons.

Stauble, D. K., and Hoel, J. (1986). Guidelines for Beach Restoration Projects, Part II -Engineering, Report SGR-77, Florida Sea Grant, University of Florida, Gainesville.

Stokes, G. G., et al. (1851). On the effect of the internal friction of fluids on the motion of pendulums. In Stokes, G. G., (ed.), *Mathematical and Physical Papers*. Cambridge: Cambridge University Press, pp. 1–10.

Suhayda, J. N. (1977). Surface waves and bottom sediment response. *Marine Geotechnology*, 2(1–4), 135–146.

Terzaghi, K. (1925). Principles of soil mechanics: III – Determination of permeability of clay. *Engineering News Record*, 95, 832–836.

Terzaghi, K. (1956). Varieties of Submarine Slope Failures. Bureau of Engineering Research Special Publication 23, University of Texas, Austin, TX.

Terzaghi, K., and Peck, R. B. (1967). *Soil Mechanics in Engineering Practice.* 2nd ed. New York: John Wiley.

Tyler, W. S. (1991). Testing sieves and their uses. *Handbook*, 53, LP9–91.

U. S. Army Corps of Engineers. (1981). Low-cost Shore Protection, Final Report on Shoreline Erosion Control Demonstration Program (Section 54 of the Water Resources Development Act of 1974 (PL 93-251)).

U.S. Army Corps of Engineers. (2008). *Coastal Engineering Manual*, EM1110-2-1100.

van der Ven, T. G. M. (1981). Effects of polymer bridging on selective shear flocculation. *Journal of Colloid and Interface Science*, 81, 290–291.

Van Olphen, H. (1977). *An Introduction to Clay Colloid Chemistry for Clay Technologists, Geologists and Soil Scientists.* 2nd ed. New York: Wiley.

Vanoni, V. A. (1975). *Sedimentation Engineering.* American Society of Civil Engineers, Manual No. 54.

Wagner, A. A. (1957). The use of the unified soil classification system by the bureau of reclamation. *Proceedings of the 4th International Conference on Soil Mechanics and Foundation Engineering* (London), 1, 125.

White, H. E., and Walton, S. F. (1937). Particle packing and particle shape. *Journal American Ceramic Society*, 20, 155–166.

Winterkorn, H. F. (1953). Macromeritic liquids. *ASTM Special Technical Publication*, 156, 77–89.

Yariv, S., and Cross, H. (1979). *Geochemistry of Colloid Systems. For Earth Scientists.* 86 figs, 32 tables, xii+450 pp. Berlin, Heidelberg, New York: Springer-Verlag.

Part IV

---

*Changing Sea Level
and Cliff Retreat*

8 Relative Sea-Level Change

8.1 INTRODUCTION

The sea level began to rise along with global temperature beginning in approximately 1900, refer to Figure 8.1 (NRC, 2012). Global eustatic sea level has risen ~120 meters since the Last Glacial Maximum (LGM; approximately 22 thousand years ago) (Khan et al., 2015; Lambeck and Chappell, 2001; Peltier, 2002; Peltier and Fairbanks, 2006; Russel, 1957). Eustatic sea level is defined as the distance from the center of the earth to the sea surface (Patzkowsky and Holland, 2012). This rise in sea level has been attributed to both natural and anthropogenic forces. The forces that contribute to the relative sea level (RSL) change are the following: (1) melting of ice, (2) changes in sea water temperature, (3) sedimentation, (4) changes in ocean boundaries, (5) interseismic and coseismic strain, and (6) salinity (Cazenave and Llovel, 2010).

The anthropogenic forcing of Earth's climate has dominated as a control of sealevel rise since approximately 1850 (Jevrejeva et al., 2009; Stammer et al., 2013). As water and ice masses spatially redistribute, Earth's crust and mantle isostatically adjust to these changes; these isostatic adjustments further contribute to local eustatic sea level (Clark and Lingle, 1979; Gehrels, 2010; King et al., 2015). These

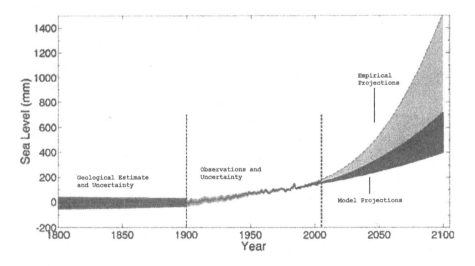

FIGURE 8.1   Observed and projected sea-level rise.

(Used with permission Board on Earth Sciences and Resources National Research Council (U.S.) Ocean Studies Board from Sea Level Rise for the coasts of California, Oregon, and Washington: Past, Present, and Future, 2012)

DOI: 10.1201/9781003454212-12

changes involve both contributions from El Nina and the melting of ice sheets. In the following, each of these contributions will be discussed.

Climate patterns such as the El Nino–Southern Oscillation affect winds and ocean circulation, raising local sea levels during warm phases (El Nino) and lowering sea levels during cool phases (La Nina) along the west coast of the United States. Large El Nino events can raise coastal sea levels by 10 to 30 cm (4 to 12 inches) for several winter months.

In addition, the large mass of glaciers and ice sheets creates a gravitational pull. This gravitational pull tends to draw the ocean water closer. As the ice melts, the gravitational pull decreases. The ice melt enters the ocean, and the land and ocean basins rise or sink (National Research Council, 2012). This vertical movement is a result of the melting land ice mass. These effects produce a distinct spatial pattern of regional sea-level change commonly referred to as a sea-level fingerprint.

An example of spatial variation of sea level is presented in Figure 8.2. A review of Figure 8.2 shows that sea levels in Juneau, Alaska have fallen. In contrast, Astoria,

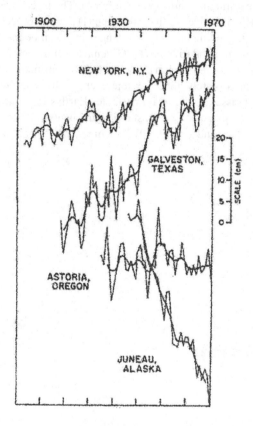

FIGURE 8.2   Yearly changes in sea levels determined from tide gauges at various coastal stations.

(From Hicks and Crosby, 1975. Courtesy of NOAA)

Oregon is shown to be relatively constant while Galveston, Texas and New York, NY are shown to increase.

8.2   HISTORICAL RELATIVE SEA LEVEL

Absolute sea level refers to the height of the ocean surface, irrespective of any movement of the land (Figure 8.3). Changes in absolute sea level are evaluated relative to a "geocentric" reference frame—i.e., relative to the center of the earth—as opposed to the elevation of the land surface. Absolute sea-level change can be combined with estimates of vertical land movement (VLM, e.g., uplift and subsidence) to project changes in "relative" sea level. RSL rise is more applicable to community planning, since ultimately it is the relative change in sea level that will determine impacts on land.

RSL change is a product both of changes in global sea level and those at the land surface. RSL is defined as the sea level that is observed with respect to a land-based reference frame (Rovere et al., 2016).

Global, or eustatic, sea level is affected by many factors. primarily ocean warming, which causes the seawater to expand, along with the melting of land/sea ice. Changes to the elevation of the land surface, or vertical land motion (VLM), are influenced by a combination of factors ranging from glaciers (due to the weight of the ice on the land), adjustments due to plate tectonics (seismic and inter-seismic rises and falls), and land subsidence (soil compaction, groundwater, and other fluid

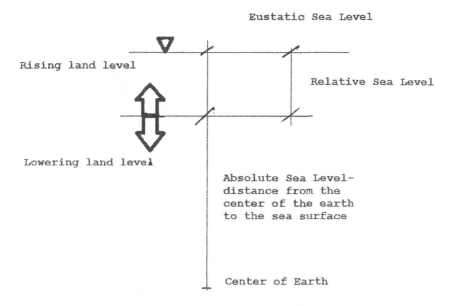

FIGURE 8.3   Illustration of relative sea level and absolute sea level.

extraction). RSL either rises if the eustatic sea level is rising faster than VLM or falls
if VLM rises faster than the sea level.

Along the Pacific northwest California coast, VLM is controlled primarily by the
Cascadia Subduction Zone (CSZ), which ranges from Cape Mendocino, CA to the
Queen Charlotte Islands in British Columbia, Canada (Figure 8.4). The CSZ is the
zone at which the Gorda and San Juan de Fuca plates are being subducted beneath

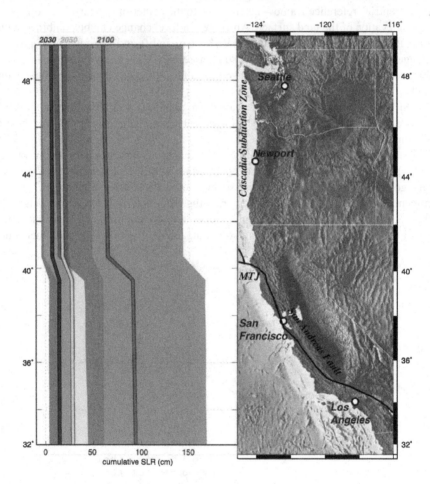

FIGURE 8.4 Projected sea-level rise off California, Oregon, and Washington for 2030
(black), 2050 (white), and 2100 (gray), relative to 2000, as a function of latitude.

Note: Solid lines are the projections, and shaded areas are the ranges. Ranges overlap, as
indicated by the dark gray shading (low end of 2100 range and high-end of 2050 range)
and middle range gray shading (low end of 2050 range and high end of 2030 range).
MTJ = Mendocino Triple Junction, where the San Andreas Fault meets the CSZ.

(Used with permission Board on Earth Sciences and Resources National Research
Council (U.S.) Ocean Studies Board from Sea Level Rise for the coasts of California,
Oregon, and Washington: Past, Present, and Future, 2012)

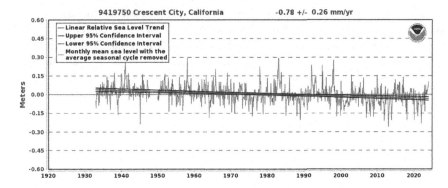

FIGURE 8.5 RSL trend, Crescent City, California.

Note: The RSL trend is −0.82 mm/yr with a 95% confidence interval of ±0.27 mm/yr based on monthly mean sea-level data from 1933 to 2022 which is equivalent to a change of −0.27 feet in 100 years.

(From NOAA, 2023. Courtesy of National Oceanic and Atmospheric Administration)

the Gorda Plate offshore Northern California and the remnants of the Juan de Fuca Plate to the north. The act of subduction results in relatively little VLM most of *the* time this is changed considerably during earthquake events. While the area between Cape Mendocino and approximately Humboldt Bay is generally rising as seismic stresses slowly relax during interseismic times, and levels near Crescent City are dropping. In Crescent City, this general negative VLM has resulted in falling sea levels in historical tide gauge levels.

Figures 8.5–8.8 presents the monthly RSL without the regular seasonal fluctuations for Crescent City, California; Charleston, Oregon; Garibaldi, Oregon; and

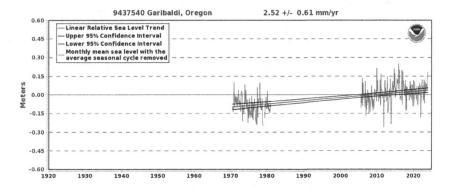

FIGURE 8.6 RSL trend, Charleston, Oregon.

Note: The RSL trend is 0.98 mm/yr with a 95% confidence interval of ±0.64 mm/yr based on monthly mean sea-level data from 1970 to 2022 which is equivalent to a change of 0.80 feet in 100 years.

(From NOAA, 2023. Courtesy of National Oceanic and Atmospheric Administration)

Friday Harbor, State of Washington. Seasonal fluctuations due to coastal ocean temperatures, salinities, winds, atmospheric pressures, and ocean currents. The RSL trend is also shown with its 95% confidence interval. RSL trends at the coast can be positive or negative. A negative trend does not mean the ocean surface is falling. It indicates the land is rising more quickly than the ocean in a particular area. Trends close to zero indicate the land is rising at nearly the same rate as the ocean.

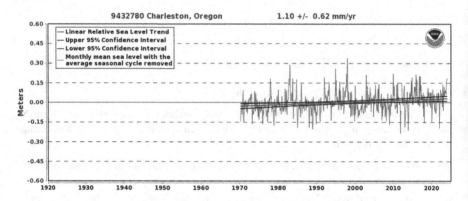

FIGURE 8.7   RSL trend, Garibaldi, Oregon.

Note: **The RSL trend is 2.44 mm/yr with a 95% confidence interval of ±0.64 mm/yr based on monthly mean sea-level data from 1970 to 2022 which is equivalent to a change of 0.80 feet in 100 years.**

(From NOAA, 2023. Courtesy of National Oceanic and Atmospheric Administration)

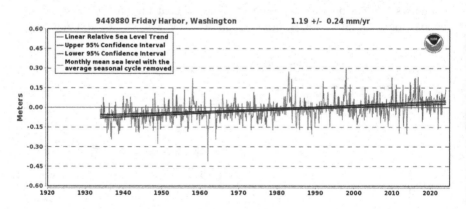

FIGURE 8.8   RSL trend, Friday Harbor, State of Washington.

Note: **The RSL trend is 1.17 mm/yr with a 95% confidence interval of ±0.24 mm/yr based on monthly mean sea-level data from 1970 to 2022 which is equivalent to a change of 0.80 feet in 100 years.**

(From NOAA, 2023. Courtesy of National Oceanic and Atmospheric Administration)

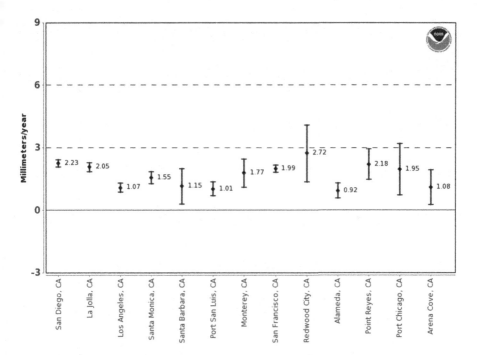

FIGURE 8.9 RSL trends for California.

(From NOAA, 2023. Courtesy of National Oceanic and Atmospheric Administration.)

As these data were collected from a land-based tide gauge, they represent RSL (i.e., The combination of both eustatic sea level and VLM). As an example over the 84 years of records, mean RSL change at the Crescent City tide gauge was -0.75 ± 0.28 mm/yr, or -0.003 ± 0.0009 ft/yr. This indicates an overall decline in RSL over the time period represented by these data.

Regional variations in RSLs for Northern California, Oregon, and the State of Washington are presented in Figures 8.9 and 8.10. These two figures compare the 95% confidence intervals of RSL regional trends. Trends with the narrowest confidence intervals are based on the longest data sets. Trends with the widest confidence intervals are based on only 30–40 years of data. The graphs give an indication of the differing rates of VLM, given that the absolute global sea level rise is believed to be 1.7 ± 0.3 mm/yr during the 20th century.

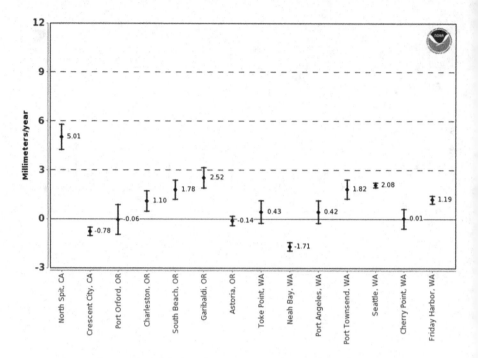

FIGURE 8.10   RSL trends for Oregon and State of Washington.
(From NOAA, 2023. Courtesy of National Oceanic and Atmospheric Administration.)

8.3   SEA-LEVEL RISE

8.3.1   GLOBAL SEA-LEVEL RISE

Based on tide gauge measurements from around the world, the Intergovernmental Panel on Climate Change (IPCC) estimated that the global sea level rose an average of about 1.7 mm/yr, over the 20th century. In contrast, global estimates based on satellite altimetry of the rate of sea-level rise range from 3.1 to 3.4 mm/yr (NRC, 2012).

8.3.2   PACIFIC NORTHWEST SEA-LEVEL RISE

8.3.2.1   General

Pacific Northwest sea-level rise has been estimated to range from 2.28 mm/yr (Burgette et al., 2009) to 2.38 mm/yr (Miller et al., 2018; Zervas et al., 2013). The National Research Council (2012) projects that the sea level will change for Washington, Oregon, and California coasts between −4 cm (sea level fall) and +23 cm by 2030, −3 cm and +98 cm by 2050, and 10–143 cm by 2100. (Cazenave and Llovel, 2010; Nerem et al., 2010; Wenzel and Schroter, 2014).

8.3.2.2 Humboldt Bay, CA

The difference between the global sea-level rise estimates and the Humboldt Bay California North Spit (NS) tide gauge observations suggests that there is subsidence of the land and the associated tide gauge. A total of 11 tidal benchmarks and associated temporary gaging stations were deployed from 1977 to 1980 when the NS tide gage was installed. Utilizing a subset of these initial observation points, current sea-level observations in Humboldt Bay have been analyzed to investigate local sea-level rise relative to regional sea-level rise. In addition, first-order leveling data has been collected by the National Geodetic Survey (NGS) to determine VLM rates for the second half of the 20th century (Burgette et al., 2009), Mitchell et al. (1994).

8.3.2.3 Crescent City, CA

The nearest NOAA continuous operating tide gauge in Crescent City (CCL) California, sea-level has observed sea-level that is lowering at 0.65 mm/yr (Zervas et al., 2013). This is the result of an upward VLM in Crescent City.

8.4 EUSTATIC SEA LEVEL

The rates of sea-level rise at specific coastal sites can differ significantly from a globally averaged rate. The extent of this regional variation in the trends of changing sea levels was shown in a review by Cazenave and Llovel (2010). This review shows nonuniform distribution of the water's temperature increase and thermal expansion is primarily responsible for spatial patterns of changing sea levels.

As these data were collected from a land-based tide gauge, they represent RSL (i.e., the combination of both eustatic sea level and VLM). Over the 84 years of records, the mean RSL change at the Crescent City tide gauge was −0.75 ± 0.28 mm/yr, or −0.003 ± 0.0009 ft/yr. This indicates an overall decline in RSL over the time period represented by these data.

8.5 PREDICTED CHANGES DUE TO EUSTATIC SEA-LEVEL RISE

The State of California Natural Resources Agency (NRC) published Sea-Level Rise Guidance ("Guidance," 2018) to provide "a bold, science-based methodology for state and local governments to analyze and assess the risks associated with sea-level rise, and to incorporate sea-level rise into their planning, permitting, and investment decisions." This Guidance provides potential rates of RSL change under a variety of scenarios. These include risk tolerance (low, medium-high, and extreme risk aversion) and greenhouse gas emissions (low and high emissions). Greenhouse gas emissions are important as the model indicates that the high emissions scenarios result in higher RSLs than the low emissions.

RSLs for Crescent City, CA were provided for the timeframe for both high and low greenhouse gas emissions from 2020 through 2100 (NRC 2018). These values are shown in Table 8.1.

From these data, the historical trend of falling RSLs caused by VLM outpacing changes in eustatic sea level is anticipated to reverse between 2020 and 2030. Based on these data, from that point through the end of the projection in 2100, RSL

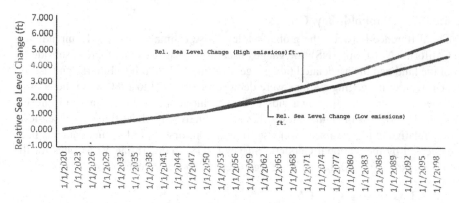

FIGURE 8.11   Projected RSL changes, 2020–2100, incorporating high and low emission scenarios.
(From NOAA, 2023. Courtesy of National Oceanic and Atmospheric Administration.)

will continue to rise. The model diverges after 2050 to reflect the potential impacts of the two different greenhouse gas emissions scenarios. Figure 8.11 illustrates the projected changes in RSL.

8.6   METEOROLOGY

Meteorology is the study of the spatial and temporal behavior of atmospheric phenomena. *Climate* using averages and various statistics characterizes the long-term meteorologic conditions of an area. Factors directly associated with climate that directly affect coastal geology such as wind, temperature, precipitation, evaporation, chemical weathering, and seawater properties. The shoreline is also affected by wave patterns due to local winds or generated by storms thousands of kilometers away. An introduction to weather patterns and coastal processes, reviewing coastal meteorology fundamentals have been provided by Hsu (1988).

8.6.1   WIND

Wind is caused by pressure gradients, which are defined as the horizontal differences in pressure across an area. Wind patterns range from global, which are typically persistent, to local and short duration, such as thunderstorms.

8.6.2   DIRECT INFLUENCE OF WIND

Wind has both a direct and indirect influence on coastline geomorphology. The direct influence includes wind as an agent of erosion and transportation. It affects the coastal zone by eroding, transporting, and subsequently depositing sediment. Bagnold (1954) found that a proportional relationship exists between wind speed and rate of sand movement. The primary method of sediment transport by wind is through saltation, or the bouncing of sediment grains across a surface. Two coastal geomorphic features that are a direct result of wind are dunes and related blowouts. *Dunes* are depositional features whose form and size are a result of sediment type, underlying topography, wind direction, duration, and strength. *Blowouts* form when wind erodes the sand from an unvegetated area leaving a depression or trough.

8.6.3 INDIRECT EFFECT

Wind indirectly affects coastal geomorphology as wind stress upon a water body causes the formation of waves and oceanic circulation.

8.6.4 LAND/SEA BREEZE

Diurnal variations in the wind result from differential heating of the ocean and land surfaces. During the day, an area of low pressure forms when the heating of the land causes the air to expand and rise. The pressure gradient between the water and the land surfaces causes a landward-directed breeze. At night, the ocean cools less rapidly than does the land, thus resulting in air rising over the ocean and subsequently seaward-directed breezes. These breezes are rarely greater than 8 m/sec (18 mph) and therefore do not have a great effect upon coastline geomorphology, although there may be some offshore-onshore transport of sediment on beaches (Komar 1998).

8.6.5 WATER LEVEL SETUP AND SETDOWN

Onshore winds cause a landward movement of the surface layers of the water and thus a seaward movement of deeper waters. Strong onshore winds, if sustained, may also cause increased water levels or setup. The opposite occurs during offshore winds.

TABLE 8.1
RSL Change, Crescent City, CA: 2020–2100

Year	Relative Sea-Level Change in Feet (Low Emissions)	Relative Sea-Level Changes in Feet (High Emissions)
2020	−0.002	−0.002
2030	0.409	0.409
2050	1.231	1.231
2070	2.400	2.800
2080	3.100	3.700
2100	4.800	5.900

8.7 SUMMARY

Sea-level change influences the entire world more than any other coastal process. The changes occurring along the shoreline of the Pacific Northwest are complicated due to the vertical movement (i.e. movement both upward and downward) of land due to a subduction zone. RSL change is defined as the difference between the sea level and the elevation of the land. Some areas along the shoreline are shown to be going up and some are going down depending on tectonics of the area. Global and slow changes especially along the north coast are always taking place, and these regional or local changes may be either rapid and/or catastrophic. Sea-level change is shown to be everywhere! The current concern about global warming and its influence on sea-level change has been a front-page story over the past century but the long-term future at any one Pacific Northwest coastal location is still a matter of speculation.

Ranging from changes on the order of meters in less than a day associated with earthquakes to only a millimeter or so in a year, sea-level changes can impact coastal management.

REFERENCES

Bagnold, R.A. (1954). Experiments on a gravity free dispersion of large solid spheres in a Newtonian fluid under shear. *Proceedings of the Royal Society of London. Series A, Mathematical and Physical Sciences*, 225(1160), 49–63.

Burgette, R. J., Weldon, R. J. III, and Schmidt, D. A. (2009). Interseismic uplift rates for western Oregon and along-strike variation in locking on the Cascadia subduction zone. *Journal of Geophysical Research*, 114(B01408), 24.

Cazenave, A., and Llovel, W. (2010). Contemporary sea level rise. *Annual Review of Marine Science*, 2, 145–173.

Clark, J. A., and Lingle, C. S. (1979). Predicted relative sea level changes (18,000 years b.P. to present) caused by late glacial retreat of the Antarctic ice sheet. *Quaternary Research*, 11(3), 279–298.

Gehrels, R. (2010). Sea-level changes since the last glacial maximum: An appraisal of the IPCC fourth assessment report. *Journal of Quaternary Science*, 25, 26–38.

Hicks, S. D., and Crosby, J. E. (1975). "An average long period sea level", series for the United States, NOAA, Tech Memorandum Nos 15: 6p.

Hsu, S. A. (1988). *Coastal meteorology*, New York: Academic Press, 260 p.

Jevrejeva, S., Grinsted, A., and Moore, J. C. (2009). Anthropogenic forcing dominates sea level rise since 1850. *Geophysical Research Letters*, 36, L20706.

Khan, N. S., Ashe, E., Shaw, T. A., Vacchi, M., Walker, J., Peltier, W. R., Kopp, R. E., and Horton (2015). Holocene relative sea-level changes from near-, intermediate-, and far-field locations. *Current Climate Change Reports*, 1, 247–262.

King, D., Schrag, D., Dadi, Z., Ye, Q., and Gosh, A. (2015). *Climate Change: A Risk Assessment*. J. Hynard and T. Rodger (eds.). Cambridge: Centre for Science and Policy. Available at: www.csap.cam.ac.uk/projects/climate-change-risk-assessment/

Lambeck, K., and Chappell, J. (2001). Sea level change through the last glacial age. *Science*, 27, 679–686.

Miller, I. M., Morgan, H., Mauger, G., Newton, T., Weldon, R., Schmidt, D., Welch, M., and Grossman, E. (2018). Projected Sea Level Rise for Washington State – A 2018 Assessment. A collaboration of Washington Sea Grant, University of Washington Climate Impacts Group, University of Oregon, University of Washington, and US Geological Survey. Prepared for the Washington Coastal Resilience Project, Updated 07/2019.

Mitchell, C. E., Vincent, P., Weldon, R. J., and Richards, M. A. (1994). Present-day vertical deformation of the Cascadia margin, Pacific Northwest, United States. *Journal of Geophysical Research Atmospheres*, 99(B6), 12,257–12,277. https://doi.org/10.1029/94JB00279

National Oceanic and Atmospheric Administration (NOAA), Sea Level Trends (2023). Tides and Currents. https://tidesandcurrents.noaa.gov/sltrends/sltrends_station.shtml?id=9410170

National Research Council (NRC) (2012). Sea-Level Rise for the Coasts of California, Oregon, and Washington: Past, Present, and Future. Committee on Sea Level Rise in California, Oregon, Washington. Board on Earth Sciences Resources Ocean Studies Board Division on Earth Life Studies: The National Academies Press. https://nap.nationalacademies.org/resource/13389/sea-level-rise-brief-final.pdf

Nerem, R. S., Chambers, D. P., Choe, C., and Mitchum, G. T. (2010). Missions. *Marine Geodesy*, 33, 435–446.

Patzkowsky, M. E., and Holland, S. M. (2012). *Stratigraphic Paleobiology*, Chicago: University of Chicago University Press.

Peltier, W. R. (2002). On eustatic sea level history: Last glacial maximum to Holocene. *Quaternary Science Reviews*, 21(1–3), 377–396.

Peltier, W. R., and Fairbanks, R. G. (2006). Global glacial ice volume and last glacial maximum duration from an extended Barbados sea level record. *Quaternary Science Reviews*, 25(23–24), 3322–3337.

Rovere, A., Stocchi, P., and Vacchi, M. (2016). Eustatic and relative sea level changes. *Current Climate Change Reports*, 2(4), 221–231. https://doi.org/10.1007/s40641-016-0045-7. S2CID 131866367.

Stammer, D., Cazenave, A., Ponte, R. M., and Tamisiea, M. E. (2013). Causes for contemporary regional sea level changes. In C. A. Carlson and S. J. Giovannoni (eds.), *Annual Review of Marine Science*, Vol. 5. Palo Alto, CA: Annual Reviews, 21–46.

Wenzel, C., Gill, S., and Schroter, J. (2013). Global and regional sea level change during the 20th century. *Journal of Geophysical Research–Oceans*, 119, 7493–7508.

Zervas, C., Gill, S., and Sweet, W. V. (2013). Estimating vertical land motion from long-term tide gauge records. *Technical Report NOS CO-OPS*, 65, 22.

9 Cliff Retreat

9.1 INTRODUCTION

The erosion of the bluffs along the shoreline is highly complex involving a number of issues. Some of these processes are wave attack, presence of a fronting beach, tides, rainfall, and groundwater seepage. In addition, it is also controlled in large part by the cliff materials and geologic structures such as bedding (Sunamura, 1992). Another factor to consider in bluff erosion is the lack of sediment supply, since currently the contribution of new sediments from rivers or ravines is interrupted by anthropic activities carried out in their basins (dams, channeling, etc.). Critical is the level achieved by the tides in combination with waves compared with the elevation of the junction between the beach and face of the sea cliff (refer to Figure 9.1). All coastlines are affected by storms and other natural events that can cause erosion. The combination of storm surge at high tide with additional effects from strong waves leads to potential bluff retreat. These conditions are commonly associated with land-falling tropical storms such as El Nino that creates the most damaging conditions.

There are several terms used to describe erosion of the cliff face. These terms are as follows:

- Abrasion: Bits of sand and larger fragments of rock that erode the shoreline or headland. The waves tend to grind down cliff surfaces like sandpaper.
- Attrition: Waves smash rocks and pebbles on the shore into each other. The rocks and pebbles are broken down and become smoother.

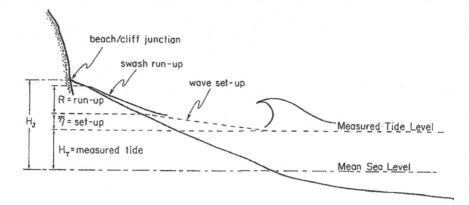

FIGURE 9.1 The basic model for the quantitative assessment of the susceptibility of sea cliffs to wave induced erosion.

(From Shih et al., 1994)

DOI: 10.1201/9781003454212-13

- Corrasion is when waves pick up beach material (e.g. pebbles) and hurl them at the base of a cliff.
- Hydraulic action occurs when waves striking a cliff face compress air in cracks on the cliff face. This exerts pressure on the surrounding rock. This pressure can progressively splinter and remove rock pieces. The cracks can grow over time, sometimes forming a cave. The rock shards fall to the sea bed where they are subjected to further wave action.

A summary of the characteristics of the entire Pacific Coast includes information on wave height (m), wave period (s), and mean tide range (m). This summary shows that in the north of Cape Mendocino, the wave height ranges from 2.2 to 2.6 m with a wave period of approximately 11 s. This corresponds to a mean range of tides from 1.5 to 2 m.

9.2 FACTORS THAT INFLUENCE EROSION RATES

9.2.1 PRIMARY FACTORS

The ability of waves to cause erosion of the cliff face depends on many factors.

- The hardness (or inversely, the erodibility) of rocks facing the sea is controlled by the rock strength and the presence of fissures, fractures, and beds of non-cohesive materials such as silt and fine sand.
- The rate at which debris from cliff erosion is removed from the foreshore depends on the power of the waves crossing the beach. This required a critical level of energy to remove material from the debris lobe. Debris lobes can be very persistent and take years to completely disappear.
- Beaches dissipate wave energy on the foreshore and provide a measure of protection to the adjoining land.
- The stability of the foreshore is its resistance to removal. Once stable, the foreshore should widen and become more effective at dissipating the wave energy. The provision of updrift material being deposited onto the foreshore beneath the cliff helps to ensure a stable beach.
- An important influence on the rate of cliff erosion is the adjacent bathymetry of the seafloor, which controls the wave energy arriving at the coast. Shoals and bars offer protection from wave erosion. This is accomplished by causing storm waves to break and dissipate their energy before reaching the shore. The changing location of shoals and bars can cause the locus of beach or cliff erosion to change position along the shore given the dynamic nature of the seafloor.
- Coastal erosion has been greatly affected by the rising sea levels globally.

9.2.2 SECONDARY FACTORS

Weathering and transport slope processes that affect cliff erosion.

- Slope hydrology
- Vegetation

- Cliff foot erosion
- Sediment accumulation at the foot of the cliff
- Resistance of sediment from the cliff foot to attrition and transport.
- Human activities

9.3   CLASSIFICATION OF COASTAL MORPHOLOGY

9.3.1   INTRODUCTION

The distribution of deltas, beaches, and spits is controlled by sediment supply and the hydrographic regime.

Geologists and coastal engineers working along the coast have long observed that the shoreline morphology of depositional coasts is strongly tied to both the wave energy and tidal currents. Waves transport sediment parallel to the coast, while tidal currents move sediment perpendicular to the shore.

A variety of coastal landforms and coastline types exists in nature. Coastal engineers are concerned with describing and classifying them. Coastal classification schemes can also be useful from a conceptual point of view. These schemes help coastal engineers to assess the different forcing factors and controls. The forcing factors and controls that produce coastal landforms are the following: sea-level history, geology, climate, waves, tides. There have been a number of approaches developed such as the following: (1) sea level approach, (2) tectonic approach, and a (3) system approach.

9.3.2   SEA LEVEL APPROACH

Early classification schemes were based on the role of sea-level variations, and thus distinguish between submerged and emerged coasts (Johnson, 1919). Submerged coasts include both drowned rivers and glacial valleys respectively. In contrast, coastal plains are characteristic of emerged coasts where sea level relative to the land has fallen. Shepard (1963, 1982) identified primary and secondary coasts. Primary coasts are the result of non-marine processes and include drowned river valleys, rocky and deltaic coasts. Secondary coasts result mainly from marine processes or organisms and include barrier coasts, coral reefs, and mangroves. An energy-based classification of coastal morphology was developed by Davies (1964) by subdividing the world's shores according to tidal range (Table 9.1). The focus on tides was

---

TABLE 9.1

Energy-Based Classification

- Microtidal 0–3 m
- Mesotidal 2–4 m
- Macrotidal > 4 m

Source: Davis (1964).

---

considered justified because tidal range controls the length of time that waves can act on any portion of the shore profile.

The chapter by Davis (1964) highlights the importance of understanding the evolution of world shorelines and the need for a comprehensive approach to studying this complex phenomenon. This system emphasizes the role of tectonic processes and sea-level changes in shaping shorelines. The chapter emphasizes the need to consider the interaction of tectonic processes in modeling changes in sea level on shoreline evolution. The aim is to provide an approach that considers various factors such as sediment supply, erosion, and deposition, to better understand and predict shoreline dynamics. The chapter introduces a morphogenic approach to understanding and modeling world shorelines, considering the interaction between tectonic processes and sea-level changes.

Hayes (1979) subsequently documented the distribution and frequency of shoreline features for each of Davies' three tidal classes, and developed morphological models for both estuaries and barrier islands (Table 9.2). This classification system was expanded by Hayes (1979) to five tidal categories for coastlines. These categories are as follows:

TABLE 9.2
Hayes (1979) Classification System

- Microtidal < 1 m
- Low-mesotidal, 1–2 m
- High-mesotidal, 2–3.5 m
- Low-macrotidal, 3.5–5 m
- Macrotidal, >5 m

The Hayes (1979) classification system was intended to be applied to trailing edge, depositional coasts. In the attempt to incorporate wave energy as a significant factor in modifying shoreline morphology the Hayes (1979) system was expanded to five shoreline categories. These categories were identified based on the relative influence of tide range versus mean wave height.

- Tide dominated (high)
- Tide dominated (low)
- Mixed energy (tide dominated)
- Mixed energy (wave dominated)
- Wave dominated

9.3.3 Tectonic Approach

Coasts can also be classified with respect to their tectonic position (Inman and Nordstrom, 1971). Leading edge coasts, also termed collision coasts, are located adjacent to subducting plate margins such as along the Pacific coasts of North and South America, Japan, and New Zealand. In the tectonic case, processes have formed mountain belts that have steep, erosive, and rocky coastlines, boulder beaches, and falling relative sea levels. Trailing edge coasts, however, are located away from subducting plate margins, are tectonically benign, older, and of lower elevation. Examples are the coasts of Africa, Australia, and the Atlantic coasts of North and South America. These coasts are typically sediment-rich, pro-gradational, with large deltas and sandy beaches.

The main shortcoming of these classifications is that they emphasize geological inheritance rather than hydrodynamic processes that shape coastal landforms. Davies (1980) identified coastal types based solely on wave height and tidal range. Because waves are generated by wind, the distribution of wave environments varies by latitude, reflecting global climate zones. Coastlines located in higher temperate and arctic latitudes are dominated by storm waves. In contrast, in lower temperate and tropical latitudes swell-dominated coasts are located.

In the middle of oceans, the tidal range is quite small (less than 1 m) but increases toward the coast and may reach in excess of 10 m. Tide amplification depends on a variety of factors. These factors are the slope and width of the continental shelf, the location and shape of continents, and the presence of large embayments. The global distribution of tidal range is controlled by large-scale coastal configuration. Macrotidal ranges exceeding 4 m are mostly observed in semi-enclosed seas and funnel-shaped entrances of estuaries. Microtidal ranges below 2 m occur along open ocean coasts and almost fully enclosed seas.

Wave height and tidal range distributions are used to infer wave- and tide-dominance of coastal processes and morphology (Davies, 1980). However, the relative effects of waves and tides are more important in shaping the coast.

If the incident wave energy is low then tide-dominated environments are not restricted to macrotidal coasts, but may also be found along microtidal coasts. There is a delicate balance between wave and tide processes for low values of wave height and tide range. These low-energy regions are shown to converge as pointed out by Davis and Hayes (1984). Therefore, tide-dominated, wave-dominated, or mixed-energy morphologies may have very little difference in wave and tide parameters.

The morphology of clastic coasts responds to the relative dominance of river outflow, waves, and tidal currents. The depositional environment of clastic coasts involves mud, sand, and/or gravel. The relative importance of these three factors is shown in a ternary diagram. Deltas are positioned at the fluvial apex because a fluvial sediment source dominates, while prograding, non-deltaic coasts are on the opposite wave–tide side, because in these environments sediment is moved onshore by waves and tides. Estuaries occupy an intermediate position because they have a mixed sediment source and are affected by river, wave, and tidal factors. Dalrymple et al. (1992) provide an evolutionary aspect by including time as a component. According to Dalrymple's classification of clastic coastal environments, time is expressed in terms of coastal accretion (or progradation) and coastal inundation (or transgression). Progradation of the coastline occurs with a relatively constant sea level and a large sediment supply. This entails the infilling of estuaries and their conversion into deltas, strand plains, or tidal flats. In contrast, transgression takes place when sea level is rising across the land, pushing the coastline farther back. Changes associated with transgression, such as the flooding of river valleys and the creation of estuaries. Dalrymple et al.'s (1992) model also acknowledges the dynamic nature of coasts, because changes in sea level lead to progradation and transgression are caused primarily by changes in climate.

9.3.4 SYSTEM APPROACH

The factors that are most important in shaping the coast have been discussed briefly above. To understand the interactions of these environmental factors a systematic approach is required. A systems approach has been developed by viewing the coast as a system with inputs and outputs of energy and material. The term "system" used here describes the framework through which energy and material are moved around the Earth. In this context there is a free exchange of energy and matter. This occurs between either different parts of the system, or between the system and its surroundings. This is called an open system. Typical open systems on the Earth's surface include river basins and beaches.

A simplified coastal system is described by Wright and Thom (1979). The movement of energy and material is presented as a process. This process represents the movement of energy from one part of the system to another. These processes can create or destroy the landforms that represent the dynamic state of the system. All systems have several shared characteristic properties. All systems are founded on the following: (1) different scales, from atomic to global; (2) they have boundaries; (3) they are associated with processes, which are the means by which changes to the system are made; and (4) they have structural organization. This latter characteristic refers to the pathways through which energy and mass are moved within the system. A system's component parts tend toward a steady or stable state over time. This final state is referred to as the system being in dynamic- or quasi-equilibrium.

The directional trajectory of processes that drive a system toward equilibrium is regulated within the system by a mechanism that is termed feedback. There are two types of feedback. Negative feedback refers to those processes which act to maintain a steady state (quasi-equilibrium) of the system. Negative feedback dampens down the system by internal self-regulation. Positive feedback refers to those processes that lead to a directional change in the system that has cumulative effect over time. Under positive feedback ("self-forcing") a system follows a directional trajectory of change away from one quasi-stable state. Negative feedback is then required to bring the system back under control and to stabilize it around a different quasi-stable state. Under some circumstances, positive feedback works so effectively that a system can accelerate out of control.

Wright and Thom (1977) introduced the term "coastal morphodynamics" to describe their approach. This term is defined as "the mutual adjustment of topography and fluid dynamics involving sediment transport." The morphodynamic approach has since become the main procedure for studying coastal evolution.

9.3.5 ENVIRONMENTAL CONDITIONS

Environmental conditions are the "set of static and dynamic factors that drive and control coastal systems" (Wright and Thom, 1977). These factors form the boundary conditions of the system, and are responsible for geographical variations in coastal geomorphology. The factors are not affected by the coastal system itself but by the main types of environmental conditions such as the geology, sediments, and external forcing.

The initial state and properties of the solid boundaries of a shoreline are described by its geology. This includes the surrounding geology and pre-existing morphologic state (shelf and shoreline configuration, lithology). Tectonics controls the width and slope of the continental shelf. Wide and flat continental shelves permit more rapid coastal progradation for a given rate of sediment supply than steep and narrow shelves. In addition, wide shelves lead to greater reduction in wave height by frictional dissipation and are also responsible for amplification of tides. Regionally, coastline configuration can also be important in controlling wave processes such as reduced wave energy levels in the lee of offshore islands and within coastal embayments. Lithology is a significant factor in the recession rate and cliff profile development along eroding rocky coasts.

Unconsolidated sediments are essential for coastal evolution because they erode into beaches. The availability of sediment depends on the location and volume of sources, and transport processes to where they are deposited. Sediments may have a marine, fluvial/deltaic, terrestrial, or biological origin. Thus they may vary in their physical, chemical, and particle size properties from place to place. Muddy sediments are most common in humid temperate or tropical climates, where they result in deltas and infilling of estuaries. Sandy sediments are characteristic of the coast and inner continental shelves in the lower midlatitudes. Gravel deposits are more common in paraglacial areas where the coastal hinterland has been glaciated.

External processes that provide the energy to drive coastal evolution are referred to in this system approach as external forcing. Important aspects of external forcing are the frequency, magnitude, and character of the energy sources. These energy sources are the following: (1) coastal winds and climate, (2) river outflow, and (3) waves, tides, and currents. A review of these energy sources indicates that the marine sources are the most important but are very dependent on atmospheric circulation. Environmental conditions change that act on the shoreline drive coastal evolution. The solid boundary conditions (i.e. rocks) operate on geological time scales and are largely controlled by vertical land movements resulting in coastal emergence or submergence. This vertical movement of the land results in falling and rising relative sea levels, respectively. On geological time scales, changing sea levels during the Quaternary affect sediment availability across the continental shelf. On a shorter time scale, human intervention in coastal catchments and drainage basins affects their ability to deliver sediment to the coast. This has significantly affected coastal sediment availability. The greatest temporal changes in external forcing, occur in seasonal or storm-related changes in weather and waves.

9.4   WAVE-INDUCED CLIFF EROSION

The mechanism of sea cliff erosion by waves is primarily controlled by two items as follows: (1) wave force exerted on the cliff and (2) resisting force of the cliff-forming materials. The forces acting on a cliff by wave action by hydraulic actions include such items as compression, tension, cavitation, and wear (Yatsu, 1966), abrasion and projectile action due to wave-carried pebbles and boulders (Bird, 1969; Cotton, 1960; Davies, 1972; Longwell et al., 1969; Thornbury, 1960; Zenkovich, 1967) and wedge action due to the air compressed in fissures by waves (Longwell et al., 1969; Thornbury, 1960).

The resisting force of the cliff-forming materials is controlled by their mechanical properties such as compressive strength and abrasive resistance (Sunamura, 1975), but also by their joint, fault, or stratification structure. The deterioration of the resisting strength against wave action is a function of (1) weathering effects (Suzuki et al., 1972) and (2) fatigue due to repeated stress application. A relationship between long-term erosion rate and the log of compressive rock strength has been presented by Sunamura (1976). The results show that the erosion rate decreases linearly with the log of increasing rock strength. In addition, chemical properties of cliff materials (such as limestone) also affect their strength.

The occurrence and amount of cliff erosion (i.e. retreat) is controlled by the relative intensity of three factors: (1) wave forces exerted on the cliff face, (2) the resisting strength of cliff materials, and (3) tides. There are also local controls on cliff erosion including the volume of sand on the fronting beach. The fronting beach acts as a buffer between the sea cliffs and storm waves (Komar and Shih, 1993; Shih and Komar, 1994).

If the wave forces are greater than the resisting strength of cliff materials then erosion and eventually cliff retreat will take place. The problem then is how to quantify these two issues and relate the erosive force to the rate of cliff erosion or retreat. There have been a number of studies which have attempted to look at this basic problem. These studies have ranged from (1) laboratory model studies, (2) mathematical models, to (3) empirical models.

9.5 LABORATORY MODEL STUDIES

Sunamura (1975) used a wave tank to study the relationship between the wave erosive force and the sea cliff erosion rate. He used a wave tank that was 25 m long, 0.6 m wide, and 0.8 m high. A flap-type wave generator was employed at one end. A model cliff was modeled having a 65-degree slope was placed at the opposite end of the tank. The cliff was constructed of a mixture of cement, fine sand, and water in a weight ratio of 1:150:50. It had a compressive strength of 340 g/cm². Waves after breaking with a period of 2.0 s and a height of 7.9 cm in simulated deep water were directed to act on the cliff. The result of the experiment showed the resulting cliff erosion and the sand beach formed in front of the cliff.

As the eroded sand was deposited at the cliff base, input waves began to use the material for abrasion. As the erosive force was increased, the erosion increased. At some point the excess deposited sand began to decrease the wave energy, so that waves gradually lost their erosive force and erosion decreased. If the eroded volume is plotted as a function of time an S-shaped curve results. These results demonstrated a feedback loop from the erosional process due to the action of sand produced by the cliff erosion process itself.

9.6   ANALYTICAL STUDIES

Sunamura (1977) modeled this process using linear automatic control theory. He proposed a relationship between $\dfrac{dX}{dt}$ = rate of horizontal erosion and the ratio of $\dfrac{f_w}{f_r}$, where f_w = wave force, and f_r = resisting strength of the cliff material. For the actual problem in the field the following equation was derived.

$$\frac{dX}{dt} \, \alpha \left[\ln\left(\frac{\rho g H}{S_c} \right) + C_3 \right], \tag{9.1}$$

where

ρ—density of water
g—gravitational acceleration
H—wave height at base of cliff
S_c—compressive strength of cliff material
$C_3 = \ln\left(\dfrac{A}{B} \right) =$ constant

9.7   FIELD STUDIES OF COASTAL RETREAT

The northern California, Oregon, and Washington coastline encompasses a wide range of coastal landforms, a product of complex geology and dynamic coastal processes. Coastal landforms include sandy and cobble beaches backed by coastal dunes, coastal bluffs, or steep cliffs. The Pacific Northwest (PNW) coastline also includes uplifted terraces, barrier spits, and estuaries and lagoons.

The majority of Northern California is highly crenulated, rocky coastline with small sections of pocket beaches with several exceptions. Near major river mouths such as the Klamath, Smith, and Eel rivers, and harbors at Eureka and Crescent City there are large beaches (Chaney, 1988; Devin, 2006). In addition, a few areas occur where steep coastal cliffs are fronted by narrow beaches. Shoreline cliff retreat rates and spatial distributions for the Northern California coastline between Patricks Point and the Oregon border are presented in Figure 9.2. A review of Figure 9.2 shows that the majority of shoreline change is accretion. This accretion occurs principally at gold Bluff beach and Dry Lagoon. Erosion occurs at Big Lagoon and the Klamath River mouth. A review of Figure 9.2 shows that Big Lagoon Beach has an erosion of approximately 2.5 m/yr while Gold Bluffs beach has an accretion of 7.2+ m/yr. The estimate for erosion at Big Lagoon Beach is in agreement with a study by Chaney (1988). The cliff erosion data along the entire California coast has been combined in a computer program called "California Cliff Erosion Viewer" which helps the user access data at specific locations (Figure 9.3). Figure 9.3 presents typical data for a location 1,600–1,646 km from the California–Mexico border.

To evaluate the historical shoreline change along the PNW coast was divided into eight regions, as shown previously in Figures 2.12 through 2.14. These eight regions are based loosely on coastal geomorphology.

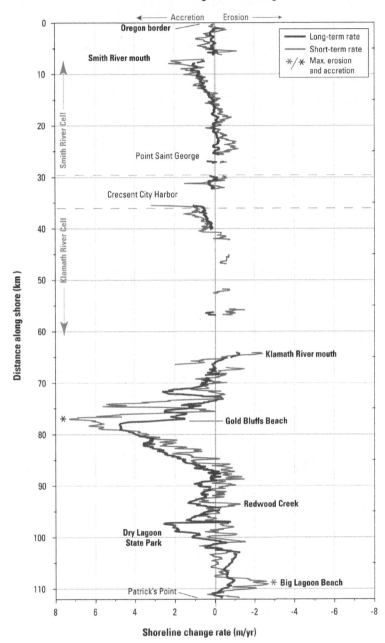

FIGURE 9.2   Shoreline change rates for the Klamath region.

Note: **The maximum long-term erosion rate was 1.2 m/yr on the south side of the Klamath River mouth, and the maximum short-term erosion rate of –2.6 m/yr was measured at Big Lagoon Park Beach.**

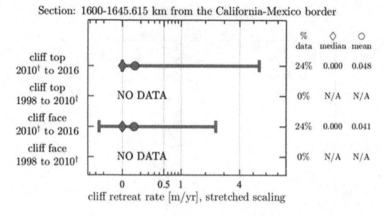

Section: 1600-1645.615 km from the California-Mexico border

cliff retreat rate [m/yr], stretched scaling

Caption: Dark gray bars indicate data range within the section. % data is the percentage of the coastal section analyzed. †The 2010 survey was conducted between 2009-2011. X-axes are square root scaled with ticks spaced at 0.1 m/yr between -1 and 1 and spaced at 1 m/yr otherwise. Data sources: Young (2018), Swirad & Young (2021, 2022a, 2022b).

FIGURE 9.3   California cliff erosion viewer for section.

(From Hapke et al., 2006)

To evaluate sediment movement gains and losses the scale was reduced to the littoral cell level as presented in Figures 9.4 through 9.11 (Rugguiero et al., 2013). These figures present both long-term (1800s through 2002) and short-term (1960s through 2002) shoreline change rates. The graphs show either accretion or erosion as a function of both spatial location and time. Figure 9.3 represents northern California from Patrick's Point to the Oregon border. Figures 9.4 through 9.10 cover the Oregon coast and Figures 9.10 and 9.11 cover the coastline of Washington State. A review of these figures shows that the average net rate of long-term change was 0.4 m/yr for Oregon and 2.2 m/yr for Washington State. In the following only the short-term changes will be discussed (i.e. from 1960s through 2002).

A review of Figure 9.4 shows that a maximum accretion of 2.5 m/yr occurred along this section of the Oregon coastline at both Gold Beach, and Cape Blanco. Maximum erosion along this same section of the coastline occurred at Otter Point (2 m/yr), Nesika Beach (2 m/yr), and north of the previous point of accretion at Cape Blanco (2.5 m/yr).

Along the section of the Oregon coast between Cape Blanco and Cape Arago (Figure 9.5) the maximum accretion occurred at Bandon (3.2 m/yr), and Cape Arago (3.2 m/yr). The maximum erosion occurred at Cape Blanco (1.5 m/yr), and just north of Bandon (2 m/yr).

Along this section of the Oregon coastline between Cape Arago and Cape Perpetua (Figure 9.6) the maximum accretion occurred south of Cape Arago (6 m/yr), Winchester Bay (6 m/yr) and Heceta Head (6.2 m/yr).

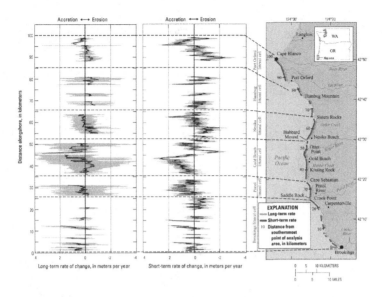

FIGURE 9.4   Graph showing long-term (1800s through 2002) and short-term (1960s through 2002) shoreline change rates (black lines on plots) for coastline between Brookings and Cape Blanco, Oregon. The location of the region is shown in small cutout. Shaded gray area behind long- and short-term rates represents uncertainty associated with the calculation.

(From Ruggiero et al., 2013)

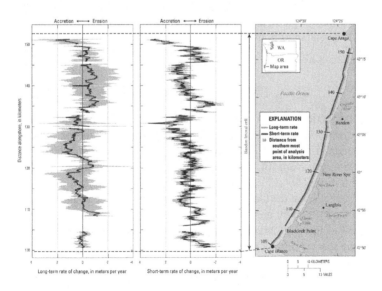

FIGURE 9.5   Graph showing long-term (1800s through 2002) and short-term (1960s through 2002) shoreline change rates (black lines on plots) for coastline between Cape Blanco and Cape Arago, Oregon. The location of the region is shown in small cutout. Shaded gray area behind long and short-term rates represents uncertainty associated with the calculation.

(From Ruggiero et al., 2013)

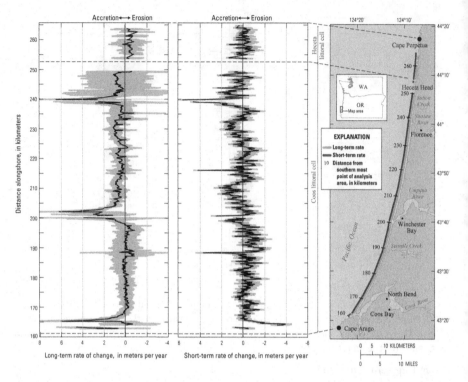

FIGURE 9.6 Graph showing long-term (1800s through 2002) and short-term (1960s through 2002) shoreline change rates (black lines on plots) for coastline between Cape Arago and Cape Perpetua, Oregon. The location of the region is shown in small cutout. Shaded gray area behind long- and short-term rates represents uncertainty associated with the calculation.

(From Ruggiero et al., 2013)

Along the section of the Oregon coastline between Cape Perpetua and Cascade Head (Figure 9.7) the maximum accretion occurred at Yaquina River (6 m/yr), Yaquina Head (3 m/yr) and Siletz Bay (3 m/yr). Maximum erosion occurred north of Yachats (2.5 m/yr), North of Yaquina John Point (2.5 m/yr), and south of Cape Foul Weather (2.5 m/yr).

Along the section of the Oregon coast between Cascade Head and Cape Falcon (i.e. Figure 9.8) the maximum accretion occurred around BayOcean (27 m/yr), and north of Nehalem Bay (6 m/yr). Maximum erosion occurred north of BayOcean (4 m/yr).

Along the section of the Oregon coast between Cape Falcon and Tillamook Head (Figure 9.9) the maximum accretion occurred a Tillamook Head (2.5 m/yr), while the maximum erosion occurred beginning at Hug Point and continuing north to south of Cannon Beach (2 m/yr).

This section (Figure 9.10) contains both the northern coast of Oregon and the coast of Washington State. The maximum accretion occurred north side of Willapa Bay (40 m/yr). The maximum erosion occurred at the south entrance to the Willapa Bay (24 m/yr).

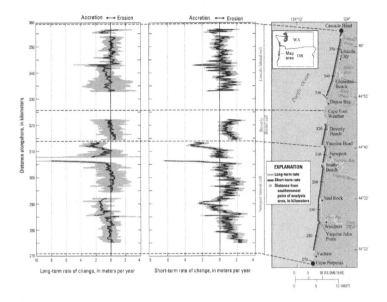

FIGURE 9.7 Graph showing long-term (1800s through 2002) and short-term (1960s through 2002) shoreline change rates (black lines on plots) for coastline between Cape Perpetua and Cascade Head, Oregon. The location of the region is shown in small cutout. Shaded gray area behind long- and short-term rates represents uncertainty associated with the calculation.

(From Ruggiero et al., 2013)

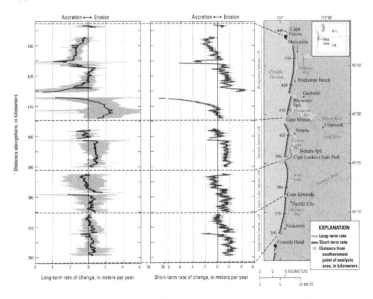

FIGURE 9.8 Graph showing long-term (1800s through 2002) and short-term (1960s through 2002) shoreline change rates (black lines on plots) for coastline between Cascade Head and Cape Falcon, Oregon. The location of the region is shown in small cutout. Shaded gray area behind long- and short-term rates represents uncertainty associated with the calculation.

(From Ruggiero et al., 2013)

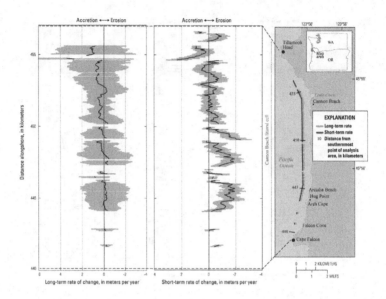

FIGURE 9.9 Graph showing long-term (1800s through 2002) and short-term (1960s through 2002) shoreline change rates (black lines on plots) for coastline between Cape Falcon and Tillamook Head, Oregon. The location of the region is shown in small cutout. Shaded gray area behind long- and short-term rates represents uncertainty associated with the calculation.

(From Ruggiero et al., 2013)

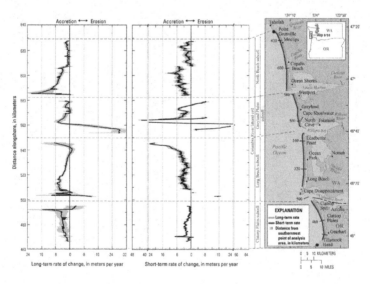

FIGURE 9.10 Graph showing long-term (1800s through 2002) and short-term (1960s through 2002) shoreline change rates (black lines on plots) for coastline between Tillamook Head, Oregon and Point Grenville, Washington State. The location of the region is shown in small cutout. Shaded gray area behind long- and short-term rates represents uncertainty associated with the calculation.

(From Ruggiero et al., 2013)

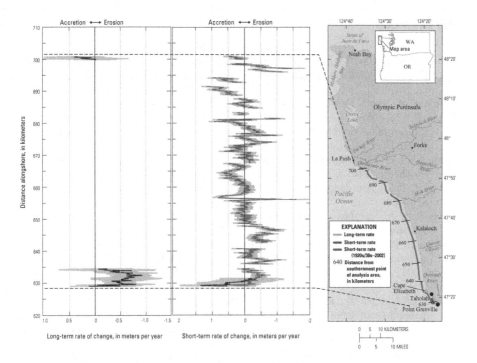

FIGURE 9.11 Graph showing long-term (1800s through 2002) and short-term (1960s through 2002) shoreline change rates (black lines on plots) for coastline between Point Grenville and Keah Bay, Washington State. The location of the region is shown in small cutout. Shaded gray area behind long- and short-term rates represents uncertainty associated with the calculation.

(From Ruggiero et al., 2013)

Along the section of the Washington State coastline (Figure 9.11) the maximum accretion occurred at Point Grenville (1.5 m/yr). In contrast, the maximum erosion occurred midway between Kalaloch and LaPush (2 m/yr)and just south of Strait of Juan de Fuca (1.5 m/yr).

9.8 SUMMARY

The erosion of the bluffs along the shoreline is highly complex involving a number of processes. Some of these processes are wave attack, presence of a fronting beach, tides, rainfall, and groundwater seepage. In addition, it is also controlled in large part by the cliff materials and geologic structures such as bedding.

The northern California, Oregon, and Washington coastline encompasses a wide range of coastal landforms, a product of complex geology and dynamic coastal processes. Coastal landforms include sandy and cobble beaches backed by coastal dunes, coastal bluffs, or steep cliffs. The PNW coastline also includes uplifted terraces, barrier spits, and estuaries and lagoons. Detailed figures are presented of the coastline from Cape Mendocino to British Columbia in which the amount of erosion and accretion are presented.

REFERENCES

Bird, E. C. F. (1969) *Coasts*, MIT Press, Cambridge, Mass: 246.

Chaney, R. C. (1988). "Coastal Bluff Retreat at Big Lagoon, California," Second International Conference on Case Histories in Geotechnical Engineering, June 1-5, St. Louis, Mo.: pp. 555–558.

Cotton, C. A. (1960). *Geomorphology*, Whitcombe and Tombs, London: 505.

Devin, S. C. (2006). "A Geomorphic Model for Coastal Cliff Recession near Crescent City, California," 40th Symposium on Engineering Geology and Geotechnical Engineering, Logan Utah, May 24–26: 12.

Davies, J. L. (1972). *Geographical Variation in Coastal Development*, Oliver and Boyd, Edinburgh: 204.

Davies, J. (1964). A morphogenic approach to worldshorelines. *Zeitschrift fur Geomorpholoty*, 8, 27–42.

Davis, R. A. Jr., and Hayes, M. O. (1984). What is a wave dominated coast? *Developments in Sedimentology*, 39, 313–329.

Dalrymple, R., Zaitlin, B., and Boyd, R. (1992). Estuarine facies models: Conceptual basis and stratigraphic implications. *Journal of Sedimentary Petrology*, 62, 1130–1146.

Davies, J. L. (1980). Geographical Variations in Coastal Development, 2nd ed. Longman, London, 212.

Hapke, C. J., Reid, D., Richmond, B. M., Ruggiero, P., and List, J. (2006). "National Assessment of Shore Line Change Part 3; Historical Shore Line Change and Associated Coastal Land Loss Along Sandy Shorelines of the California Coast," USGS Open-File Report 2006–1219: 72.

Hayes, M. O. (1979). Barrier island morphology as a function of wave and tide regime, in Leatherman, S. P. ed. *Barrier Islands from the Gulf of St. Lawrence to the Gulf of Mexico*, New York, NY: Academic Press, 1–29.

Inman, D. L., and Nordstrom, C. E. (1971). On the tectonic and morphologic classification of coasts. *Journal of Geology*, 79(1), 1–21.

Johnson, D. W. (1919). *Shore Processes and Shoreline Development*, New York: John Wiley & Sons.

Komar, P. D., and Shih, S.-M. (1993). Cliff erosion along the Oregon coast: A tectonic-sea level imprint plus local controls by beach processes. *Journal of Coastal Research*, 9(3), 747–765.

Longwell, C. R., Flint, R. F., and Sanders, J. E. (1969). *Physical Geology*, New York: John Wiley and Sons.

Ruggiero, P., Kratzmann, M. G., Himmelstoss, E. A., Reid, D., Allan, J., and Kaminsky, G. (2013). National assessment of shoreline change: Historical shoreline change along the Pacific Northwest coast. U.S. Geological Survey Open-File Report 2012–1007, 62 p. http://dx.doi.org/10.3133/ofr20121007.

Shih, S.-M., and Komar, P. D. (1994). Sediments, beach morphology and sea cliff erosion with an Oregon coast littoral cell. *Journal of Coastal Research*, 10, 144–157.

Shih, S. H., Komar, P. D., Tillotson, K. J., McDougal, W. G., and Ruggiero, P. (1994). Wave runup and sea cliff erosion, *Proceedings Coastal Engineering 1994, ASCE*, 2170–2184.

Shepard, F. P., Goldberg, E. D., and Inman, D. L. (1963). *Submarine Geology*, Harper Publication.

Shepard, F. P. (1982). North America, coastal morphology. In Schwartz, Maurice L. (Ed.), *Encyclopedia of Beaches and Coastal Environments, Encyclopedia of Earth Sciences*, Volume XV, Hutchinson Ross Publishing Co., 940.

Sunamura, T. (1975). A laboratory study of wave-cut platform formation. *Journal of Geology*, 83, 389–397.

Sunamura, T. (1976). Feedback relationship in wave erosion of laboratory rocky coast. *Journal of Geology*, 84, 427–437.

Sunamura, T. (1977). A relationship between wave-induced cliff erosion and erosive force of waves. *Journal of Geology*, 85, 613–618.

Sunamura, T. (1992). Coastal cliff erosion due to waves: field investigations and laboratory experiments, Geomphology of Rocky Coasts. Wiley

Suzuki, T., Takahashi, K., and Sunamura, T. (1972). Rock control in coastal erosion at Arasaki, Miura Peninsula, Japan (Abs.), International Geog. Congress, 22nd, Canada: 66–68.

Thornbury, W. D. (1960). *Principles of Geomorphology*, New York: John Wiley and Sons.

Wright, L. D., and Thom, B. (1977). Coastal depositional landforms: A morphodynamic approach. *Progress in Physical Geography*, 1, 412–459.

Yatsu, E. (1966). *Rock Control in Geomorphology*, Tokyo: Sozosha.

Zenkovich, V. P. (1967). *Processes of Coastal Development*, Edinburgh: Oliver and Boyd.

Part V

Effects on Man-Made Structures

10 Cliff Retreat Mitigation

10.1 INTRODUCTION

Coastline protection systems typically include both man-made structural (i.e. hard engineering) and natural habitats (i.e. soft engineering) features. The relationships and interactions among these features are important variables in determining coastal vulnerability, reliability, risk, and resilience.

Coastal risk reduction can be achieved through several approaches. These features may be used in combination with each other. Options available to reduce coastal risk include (1) natural or nature-based measures, (2) structural measures, and (3) non-structural measures. The types of risk reduction measures employed in any project depend upon the geological setting, the desired level of risk reduction, objectives, cost, reliability, and other factors.

Hard-erosion control methods provide a more permanent solution than soft-erosion control methods. Hard erosion control methods include seawalls, revetments, and groins which serve as semi-permanent infrastructure. These structures are not immune from normal wear-and-tear and will have to be periodicity refurbished or rebuilt. It is estimated the average life span of a seawall is 50–100 years and the average for a groyne is 30–40 years. Because of their relative permanence, it is normally assumed that these structures can be a final solution to erosion.

Seawalls can also restrict public access to the beach and drastically alter the natural state of the beach. Groynes also drastically alter the natural state of the beach. Other criticisms of seawalls and groynes are that they can be expensive, difficult to maintain, and can sometimes cause further damage to the beach if built improperly. Alternatives to coastal erosion mitigation are presented in Table 10.1 (French, 2001). Estimated costs associated with both hard and soft solutions to shoreline protection are presented in Table 10.2.

10.2 HARD ENGINEERING COASTAL PROTECTION

10.2.1 INTRODUCTION

The purpose of traditional hard engineering coastal protection strategies is to slow down or eliminate erosion of the shoreline. This is normally accomplished by placing an artificial, resistant barrier between the wave action and the shoreline. Seawalls and revetments differ in their engineering function, but often have similar construction details. Seawalls and revetments can be classified as either sloping-front or vertical-front structures. Sloping front structures may be constructed as flexible rubble-mound structures which are able to adjust to some toe and crest erosion.

DOI: 10.1201/9781003454212-15

TABLE 10.1
Alternatives for Coastal Hazard Mitigation

Armoring structures	Seawall
	Bulkhead
	Revetment
Beach stabilization	Breakwaters (including artificial headlands)
	Groins
	Sills
	Vegetation
	Groundwater drainage
Beach restoration	Beach nourishment
	Sand passing
Adaption and accommodation	Flood proofing
	Zoning
	Retreat
Combinations	Structural and restoration
	Structural and restoration and adaption
Do nothing	No intervention

Source: Adaption from CEM Part V, Table V-3-1.

TABLE 10.2
Estimated Cost Associated with Hard and Soft Shoreline Erosion Protection

Coastal Structure	Application	Technique	Initial Construction Cost (plf)	Operation and Maintenance (plf)
Revetment	Large wave	Coastal	$5001–$10,000	$101–$500
	Large fetch	Structure		
	Steep slope			
	Open coast			
Bulkhead	Large wave	Coastal	$2001–1001	$1001–2000
	Large fetch	Structure		
	Steep slope			
	Open coast			
Seawall	Large wave	Coastal	$5001–$10,000	$2001–$5000
	Large fetch	Structure		
	Steep slope			
	Open coast			
Breakwater	Large wave	Coastal	$5001–$10,000	$2001–$5000
	Large fetch	Structure		
	Steep slope			
	Open coast			

(Continued)

TABLE 10.2 *(Continued)*
Estimated Cost Associated with Hard and Soft Shoreline Erosion Protection

Coastal Structure	Application	Technique	Initial Construction Cost (plf)	Operation and Maintenance (plf)
Groin	Large wave Large fetch Steep slope Open coast	Coastal Structure	$2001–$5000	$1001–$2000
Beach Nourishment	Small wave Small fetch Gentle slope Sheltered coast	Living Shoreline	$2001–$5000	$1001–$2000
Beach Nourishment And Vegetation on dune	Small wave Small fetch Gentle slope Sheltered coast	Living Shoreline	$2001–5000	$1001–2000
Vegetation Only	Small wave Small fetch Gentle slope Sheltered coast	Living Shoreline	Up to $1000	Up to $1000
Edging	Small wave Small fetch Gentle slope Sheltered coast	Living Shoreline	$1001–$2000	Up to $1000
Sills	Small wave Small fetch Gentle slope Sheltered coast	Living Shoreline	$1001–$2000	Up to $1000

Notes: plf – per linear foot.
Source: Adapted from Coast.NOAA.gov (2024).

Figure 10.1 shows three examples with randomly placed armor (Corps of Engineers, 1953, 1984, 2003).

The concerns about using hard engineering techniques to protect the coast go beyond cost and extend to their interruption of natural systems at the coast. The input of sediment into the system is reduced by slowing down or preventing coastal erosion. This has implications in turn for beach size, deposition, and transfers of sediment to neighboring properties.

10.2.2 REVETMENT

A revetment is defined as a facing used to support an embankment. Figure 10.1 shows three examples of revetments with armor.

10.2.2.1 Benefits

Revetments can be open slanted concrete, geosynthetic or wooden facing/fence offering partial resistance but letting some seawater pass through (refer to Figure 10.1) (Ghiassian et al., 1997). A revetment is a passive structure, which protects against erosion caused by wave action, storm surge, and currents. The main difference in the function of a seawall and a revetment is that a seawall protects against erosion and flooding, whereas a revetment only protects against erosion.

It is also cheaper to construct a revetment than a sea wall along a beach. In addition, a revetment allows beach material to be deposited behind. It also reduces the power of oncoming waves.

10.2.2.2 Concerns

A seawall and groins can restrict access to the sea from a beach and are unattractive along a length of beach. They can also be damaged in high energy conditions and

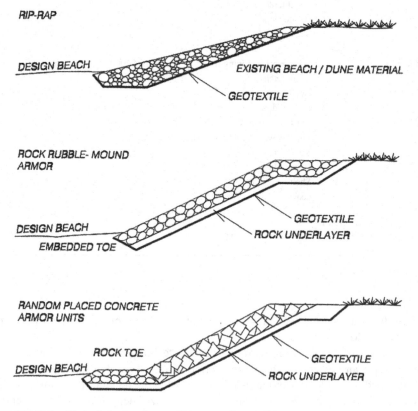

FIGURE 10.1 Typical sloping front rubble mound seawall/revetment structures.

(From Corps of Engineers, 2003)

require regular maintenance and repair. Revetments have many of the same issues as seawalls.

The stability of the slope whether rubble mound or concrete armor units is very dependent on an intact toe support. The loss of toe support will likely result in either significant armor layer damage, or complete failure of the armored slope.

These strategies along with a discussion of suitability, construction material, benefits, and concerns are summarized below (Coast.NOAA.gov, 2024; Hall and Pilkey, 1991).

10.2.3 SEA WALL

A seawall is defined as an embankment to prevent erosion of a shoreline (refer to Figure 10.2). The sea wall is any structure separating land and water areas (refer to Figure 10.3). They are designed to prevent coastal erosion and other damage due to wave action and storm surge such as flooding.

10.2.3.1 Suitability

Sea walls need to be built on a rigid foundation.

10.2.3.2 Material Options

Concrete, gabion, rip-rap, or a combination, refer to Figure 10.3.

FIGURE 10.2 Concrete sea wall.

(From Wikipedia, Creative Commons CC0 License)

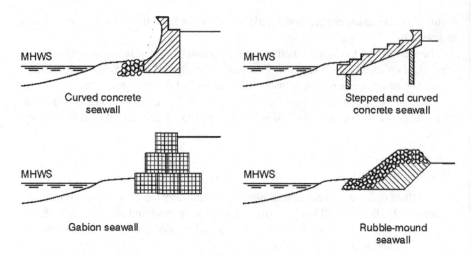

FIGURE 10.3 Various types of seawalls.

(From Coastal Wiki, 2023; see https://www.coastalwiki.org/wiki/Seawall)

10.2.3.3 Benefits

- Traditional, long used, and with proven effectiveness.
- Absorbs and deflects wave energy back to sea. If equipped with recurved upper lip dampens down oncoming wave power.
- The estimated life span of a seawall is 50–100 years.

10.2.3.4 Issues

- Requires regular repair to prevent quarrying at base. If not repaired can undermine sea wall foundations.
- Expensive to construct and maintain.
- Gives an artificial appearance to the coast.
- Seawalls can deprive public access to the beach and drastically alter the natural state of the beach.
- Seawalls tend to interfere with the natural water currents and prevent sand from shifting along coasts.
- Seawalls have a tendency to cause erosion in adjacent beaches and dunes.
- Unintended diversion of stormwater and into other properties.
- Hard erosion control structural solutions tend to cause more problems than they solve.

10.2.4 RIP RAP/ROCK ARMOR

Massive blocks of natural rock placed in position and piled up at the base of a cliff. Riprap describes a range of rocky material placed along shorelines, bridge foundations, steep slopes, and other shoreline structures to protect from scour and erosion.

10.2.4.1 Suitability

Rip/rock armour can be placed on a subgrade surface on which the rock riprap is to be placed should be cut or filled and graded to the lines and grades shown on design drawings.

10.2.4.2 Material Options

Crystalline rock. Rock rip rap should be dense, sound and free from cracks, seams and other defects conducive to accelerated weathering. The rock should be angular to sub-rounded in shape with the greatest dimension not greater than 2 times the least dimension. It should be free from dirt, clay, sand, rock fines, and other material not meeting the required gradation limits. Rock hardness shall be such that it will not dent when struck with the rounded end of a one pound ball peen hammer, or hardness shall be determined by other methods.

10.2.4.3 Benefits

Requires less maintenance than a sea wall. Granite is often used that is barely eroded even under the highest energy conditions. May look more natural than a concrete sea wall.

10.2.4.4 Concerns

Crystalline rock is expensive to extract, transport, and place in position. The rock can impede access to a beach by visitors. This can lead to injuries to individuals climbing over. Rodents may inhabit spaces between rocks.

10.2.5 TETRAPODS

Tetrapods are molded multi-angular concrete shapes formed on-site and tipped onto a beach to form interlocking components (refer to Figure 10.4). A tetrapod is a form

FIGURE 10.4   Tetrapods used as sea wall.

(From Wikipedia, Creative Commons CC0 License)

of wave-dissipating concrete block used to prevent erosion caused by weather and longshore drift, It is used primarily to reinforce coastal structures such as seawalls and breakwaters (Komar, 1983). Tetrapods use a tetrahedral shape to dissipate the force of incoming waves by allowing water to flow around rather than against them, and to reduce displacement by interlocking.

10.2.5.1 Suitability
Tetrapods can be placed on a prepared level subgrade surface that should be cut or filled to form a foundation using geotextiles overlying rubble armor stones.

10.2.5.2 Material Options
Cast reinforced concrete.

10.2.5.3 Benefits
Cheaper than rock armor but doing the same role but are constructed on site from concrete. Effective along long stretches of coastline requiring protection.

10.2.5.4 Concerns
Tetrapods are less attractive than natural rock— they look artificial. May protrude into sea and endanger swimmers/small craft Almost impossible to climb over to get access to a beach

10.2.6 GABIONS

Gabions are rock-filled steel wire cages placed along an eroding vulnerable coast.

10.2.6.1 Suitability
They function by absorbing the energy of waves. This allows the build-up of a beach. They can be expensive to obtain, place, and transport the necessary rock boulders.

10.2.6.2 Material Options
Any erosion resistance crystalline rock may be used.

10.2.6.3 Benefits
Cheaper than tetrapods, but doing the same role and may look more attractive than alternatives.

10.2.6.4 Concerns
Wire containers may rust and be broken under high energy conditions. Requires regular repair & replacement. Rodents may inhabit spaces between rocks.

10.2.7 GROINS (I.E. GROYNES)

Groins are a rigid hydraulic structure typically constructed at right angles to a beach extending into the sea (refer to Figures 10.5 and 10.6). They are designed to stabilize a stretch of beach against erosion due primarily to longshore currents. Drift sediments are used to build up beach width and height (Gaughan and Komar, 1977).

FIGURE 10.5 Groins employed to retain beach.

(From Wikipedia, Creative Commons CC0 License)

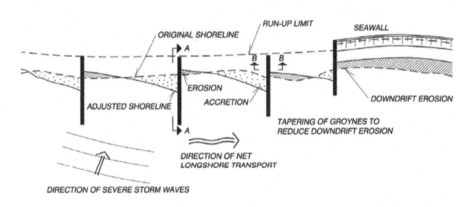

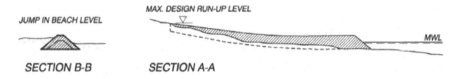

FIGURE 10.6 Typical beach configuration utilizing groins and illustrating erosion and accretion issues.

(From Corps of Engineers, 2003)

10.2.7.1   Benefits

1. Utilized to build up or widen a beach by trapping the longshore drift
2. Stabilize a beach that is subject to severe storms
3. Reduces the rate of sand loss by longshore transport
4. Reduces the rate of longshore transport out of area by locally restoring shore-line so that it is nearly parallel with the predominant incoming wave crests.
5. Reduces longshore losses of sand from an area by compartmenting the beach
6. Prevent sedimentation or accretion in a downcoast area (i.e. inlet) by acting as a barrier (Komar, 1998).

10.2.7.2   Issues

Traditionally constructed of hardwood—which is increasingly environmentally unsustainable. Requires maintenance and repair. Speeds up downcoast erosion by robbing adjacent beaches of sand as illustrated in Figure 10.6.

10.2.8   OFFSHORE REEFS

Artificial sand/gravel offshore deposits designed to intercept destructive wave action. Tombolos are more likely to form when breakwaters are constructed within the surf zone. The two examples of detached breakwaters shown in Figure 10.7 serve different functions.

10.2.8.1   Suitability

The most "natural" of hard engineering techniques. Creates additional shore habitat in calmer water conditions between the offshore reef and shore—benefits tourist use.

10.2.8.2   Material Options

Gravel and sand.

10.2.8.3   Benefits

Effective at increasing a natural barrier of beach between sea and shore. Tourism amenities such as wider beaches attract more visitors. Attractive Groines can act as "wind breaks" for visitors. Calmer inshore water.

10.2.8.4   Issues

Vulnerable to storm conditions. Less reliable than "concrete/rock" strategies but may be overwhelmed by rising sea levels.

10.2.9   DETACHED BREAKWATERS

Detached breakwaters are almost always built as rubble mound structures. Typical beach configurations with detached nearshore breakwaters are shown in Figure 10.7. Whether or not the detached breakwaters become attached to shore is a function of placement distance offshore. Tombolos are more likely to form when breakwaters are constructed within the surf zone. The two examples of detached breakwaters shown in Figure 10.7 serve different functions. Functional design guidance on detached break-waters and rubble mound groin is presented in part 6 of Corps of Engineers (2003).

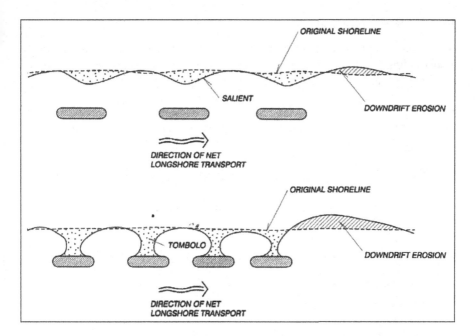

FIGURE 10.7 Detached breakwaters.

(From Corps of Engineers, 2003)

10.3 SOFT ENGINEERING COASTAL PROTECTION

10.3.1 INTRODUCTION

In the last few decades there has been a trend away from hard to soft engineering in coastal erosion control. Soft construction for coastal protection tries to use naturally available materials. The advantage of the soft engineering approach is both the aesthetic appearance of the natural materials employed and their compatibility with natural coastal processes. The use of soft engineering approaches tends to reduce the initial cost, but there are sometimes issues with the long-term viability of projects.

Soft engineering approaches are sometimes referred to as living shorelines which uses plants and other natural elements. The living shorelines have been found to be more resilient against storms improve water quality, increase biodiversity, and provide fishery habitats. Marshes and oyster reefs are examples of vegetation that can be used for living shorelines, they act as natural barriers to waves. Typically, fifteen feet of marsh can absorb fifty percent of the energy of incoming waves.

In a study by Gittman et al. (2014) marshes were found to protect shorelines from erosion better than bulkheads (i.e. seawalls). They found that Hurricane Irene damaged 76% of bulkheads surveyed, while very little or no damage was found in marshes protected by living shorelines.

Living shorelines provide a softer shoreline that is more beautiful and resilient than a bulkhead. They are also much cheaper to install and maintain than hard structures such as bulkheads and seawalls. Living shorelines are often more effective if they stretch along an entire beach rather than just one property.

10.3.2 BEACH NOURISHMENT

Beach nourishment involves placing a large volume of sand from an outside source to an eroding beach. The addition of sand widens the beach and tends to move the shoreline seaward.

10.3.2.1 Suitability

This procedure can be employed on low-lying oceanfront areas with existing sources of sand.

10.3.2.2 Benefits

- Expands usable beach area.
- Lower environmental impact as compared to hard structures.
- Flexible strategy.
- Project can be redesigned with relative ease.
- Provides habitat and ecosystem services.

10.3.2.3 Disadvantages

- Requires continual sand resources for renourishment.
- No high water protection.
- Appropriate in limited situations.
- Possible impacts to regional sediment transport.

10.3.2.4 Comments

The placement of fill sand on beaches that are eroding was one of the first approaches in soft construction. Beach fills have two major functions: (1) to provide temporary protection to upland property, and (2) to increase temporarily the recreational space along the shore. Neither of these functions can be satisfied if the sediment used for filling is either finer or coarser than the existing sand. This is because the recreational function is typically reduced by using material coarser than sand. Therefore, the primary beach fill material is usually sand.

The sediment employed for a beach fill is known as the borrow material. In contrast, the existing sediment on the beach prior to the fill is known as the native material. The sediment property important for beach nourishment design is the particle size distribution of the borrow and native sands. The beach fill design consists of calculating the volume of borrow material required with a given size distribution that will produce a required volume of beach fill.

The median particle size of the borrow sand ideally should not be less than the median size of the existing native sand. In addition, the range of particle sizes in the borrow particle size distribution should not exceed the spread of sizes in the native sand. Often it is not possible to meet these ideal conditions either because suitable borrow material does not exist in adequate volume or at a reasonable cost. Further, on severely eroded beaches, the native sand may be skewed to coarser size ranges because the fines have been eroded out. Thus producing unrealistic requirements for a borrow sand particle size distribution.

Coastal engineers design the size and shape of the beach to be compatible with the following: (1) the adjacent bathymetry, (2) available particle size distribution, and (3) wave conditions. As indicated before borrow material is selected to have either the same size distribution or have larger particle sizes than the native beach material and without mud. In addition, the borrow site must be located where removal of material will cause minimal damage to the existing benthic environments.

The purpose of beach fill design is to compensate for the differences between borrow sand and native sand. This is typically accomplished by increasing the amount of borrow sand and assuming a percentage loss of the fine fractions.

This occurs during the material handling process between borrow and beach. There have been cases where such handling losses have produced sand fill on the beach that is coarser than the borrow sand from which the fill was derived. Usually, material comprising a beach provides more protection against erosion when its particles are coarser and angular.

The shore protection and the recreational qualities of a beach will conflict when coarser sediment sizes are used. In addition, odor and color of beach fill may also be objectional to recreational users, but experience has shown that, if the grain size of the material is appropriate, the objectional odor and color are normally temporary.

10.3.3 DUNES

A dune is defined as a hill or ridge of wind-blown sand. The dune exists because dry sand and various wind conditions are common along the back beach environment. The back beach environment is rarely wet and generally devoid of vegetation. Wet sand exhibits apparent cohesion due to the effect of surface tension. The dry back beach typically shows various signs of wind transport, including ripples, sand shadows, and heavy minerals or gravelly concentrations of shells and shell debris. The sand shadows indicate a recent wind direction and may show scour around a shell or pebble.

The gravel or shell lag deposit results from wind blowing the fine sand from the beach and leaving the larger particles that cannot be transported. A pavement of large particles can almost form after a sustained wind. Such a pavement inhibits further wind erosion (Bagnold, 2005).

The wind-blown beach sand tends to accumulate just landward of an active backbeach. It is stopped from further transport by any type of obstruction that may be present, including bedrock cliffs, vegetation, existing dunes, or sea walls. Once the initiation of eolian sediment accumulation begins, it continues unless conditions change. These types of conditions are (1) loss of sediment supply, (2) the destruction of the stabilizing factor, or (3) wave-induced erosion.

Vegetation is one of the most widespread aids in controlling dune development. Any type of plant serves as a focus for anchoring wind-blown sediment. Typically the back beach is covered with opportunistic plants surrounded by small piles of sand. As the sand becomes trapped by the vegetation this provides an enlarged area of stability. As additional sand becomes trapped a small dune will develop. Under proper conditions, a dune will eventually become larger and coalesce into a foredune ridge.

Small incipient dunes are quite vulnerable to even a modest storm. This is the primary reason that attention is paid to preserving vegetation on the back beach and at the foot of the dunes. Dune size is dependent largely on the supply of sand-sized sediment.

The presence of dunes is evidence of the mobility of sand through wind transport on the coast. In contrast, the attack by waves is a major factor in the stability of dunes. Although vegetation is an effective stabilizer of sand there are conditions when even vegetated dunes may become either mobile or eroded. Dunes are quite vulnerable to wave attacks produced by storms. In areas of generally erosive beach conditions dune retreat is a problem because there is no back beach. Elevated water level with superimposed storm waves produces swash and, in some cases, direct wave attack at the toe of the dune. The sand is easily washed away and carried both offshore and alongshore. Even though a dense dune grass cover typically is present, the sand is easily removed, commonly leaving a dense root system hanging over the scarp in the dune. Post-storm recovery may occur and return some, or even all, of the sand to the beach, but it may take a number of years to restore the loss caused by just a single storm. Rising sea level presents another scenario for dune erosion, by providing a continual increase in the accessibility of the dunes to wave attack.

Another major aspect of dune dynamics is concerned with the migration of part or all of the dune through eolian processes. The same mechanisms that form the dune also can cause it to move, sometimes great distances. The most common process for a migration is called blowover. The onshore wind component simply carries sand across the dune surface and permits it to move down the landward side by gravity.

10.3.4   Utilization of Plant Material

10.3.4.1   Living shoreline with only vegetation or in combination with beach nourishment with vegetation

Beach nourishment with vegetation on an accompanying dune to help anchor sand and provide buffer to protect inland areas from waves, flooding, and erosion.

10.3.4.1.1   Suitability

Low-lying oceanfront areas with existing sources of sand and sediment.

10.3.4.1.2   Material Options

Sand with vegetation. This combination can strengthen dunes with geotextile tubes and/or use a rocky core in dune.

10.3.4.1.3   Benefits

- Expands usable beach area.
- Lower environmental impact.
- Flexible strategy.
- Redesigned with relative ease.
- Vegetation strengthens dunes and increases their resilience to storm events.
- Provides habitat and ecosystem services.

10.3.4.1.4 *Disadvantages*

- Requires continual sand resources for renourishment.
- No high water protection.
- Appropriate in limited situations.
- Possible impacts to regional sediment transport.

10.3.4.2 Edging

Structure to hold the toe of existing or vegetated slope in place. Protects against shoreline erosion.

1. Suitability
   - Most areas except high wave energy environments
2. Material options
   - Stone, living reef (ocyster/mussel), gabions
3. Benefits
   - Provides habitat and ecosystem service
   - Dissipates wave energy
4. Disadvantages
   - No storm surge reduction ability.
   - No high water protection.
   - Appropriate in limited situations.
   - Uncertainty of successful vegetation growth and competition with invasive species.

10.3.4.3 Sills

A marsh sill is a type of low-profile stone structure used to contain sand fill and create a newly planted area that dissipates wave energy. In addition, the marsh can help reduce erosion farther inland. A gapped approach to the stone structure allows habitat connectivity, greater tidal exchange, and better waterfront access.

10.3.4.3.1 *Suitability*

- Most areas except high wave energy environments

10.3.4.3.2 *Vegetation Base With Material Options*

- Stone
- Sand breakwaters
- Living reef (oyster/mussel)
- Rock gabion baskets

10.3.4.3.3 *Benefits*

- Provides habitat and ecosystem service
- Dissipates wave energy
- Slows inland and ecosystem services
- Increases natural stormwater infiltration
- Toe protection helps prevent wetland edge loss

10.3.4.3.4   Disadvantages

* Require more land area
* No high-water protection
* Uncertainty of successful vegetation growth and competition with invasive species

10.4   SUMMARY

Coastal systems typically include both man-made structural (i.e. hard engineering) and natural habitats (i.e. soft engineering) features. The relationships and interactions among these features are important variables in determining coastal vulnerability, reliability, risk, and resilience.

Hard-erosion control methods provide a more permanent solution than soft-erosion control methods. Hard erosion control methods include seawalls, revetments, and groins which serve as semi-permanent infrastructure. These structures are not immune from normal wear-and-tear and will have to be periodicity refurbished or rebuilt.

Soft engineering approach or sometimes referred to as living shorelines uses plants and other natural elements. The living shorelines have been found to be more resilient against storms improve water quality, increase biodiversity, and provide fishery habitats. The soft engineering approaches are limited in their ability to deal with high water surges and high waves.

REFERENCES

Bagnold, R. A. (2005). *The Physics of Blown and Desert Dunes*, Dover Earth Science: 336.

Coast.noaa.gov (2024). Natural and structural measures for shoreline stabilization. http://coast.noaa.gov/data/digitalcoast/pdf/living-shoreline.pdf

Coastal wiki (2023). http://coastwiki.org/wiki/seawall

Corps of Engineers (1953). Shore Protection Planning and Design, The Bulletin of the Beach Erosion Board, Special Issue No.2, U.S. Army, March: 360.

Corps of Engineers (1984). Shore Protection Manual, Vol 1 and 2, U.S. Army.

Corps of Engineers (2003). *Coastal Engineering Manual*, EM1110-2-1100, Parts 1 thru 6, U.S. Army.

French, P. W. (2001). *Coastal Defences: Processes, Problems, Solutions*, Routledge: 366.

Gaughan, M. K., and Komar, P. D. (1977). Groin length and the generation of edge waves. Proceedings 15th Coastal Engineering Conference, American Society of Civil Engineers: 1459–76.

Ghiassian, H., Gray, D. H., and Hryciw, R. D. (1997). Stabilisation of coastal slopes by anchored geosynthetic systems. *Journal of Geotechnical and Geoenvironmental Engineering*, 123(8), 736–43.

Gittman, R. K., Popowich, A. M., Bruno, J. F., and Peterson, C. H. (2014). Marshes with and without sills protect estuarine shorelines from erosion better than bulkheads during a category 1 hurricane. *Ocean and Coastal Management*, 102, 94–102.

Hall, M. J., and Pilkey, O. H. (1991). Effects of hard stabilisation on dry beach width for new Jersey. *Journal of Coastal Research*, 7(3), 771–85.

Komar, P. D. (1983). Coastal erosion in response to construction of jetties and break-waters. In Komar, P. D. (Ed.), *Handbook of Coastal Protection and Erosion*, 191–204. Boca Raton, FL: CRC Press.

Komar, P. D. (1998). *Beach Processes and Sedimentation*, 2nd ed. Englewood Cliffs, NJ: Prentice Hall.

Index

Note: Locators in *italics* represent figures and **bold** indicate tables in the text.

Printed in the United States
by Baker & Taylor Publisher Services